Coastline Changes

Coastline Changes

A Global Review

Eric C. F. Bird
University of Melbourne
and
United Nations University

A Wiley-Interscience Publication

JOHN WILEY & SONS
Chichester · New York · Brisbane · Toronto · Singapore

Copyright © 1985 by John Wiley & Sons Ltd

Library of Congress Cataloging in Publication Data:
Bird, E. C. F. (Eric Charles Frederick), 1930–
  Coastline changes.

  'A Wiley–Interscience publication.'
  Bibliography: p.
  Includes index.
  1. Coast changes.   I. Title.
GB451.B513 1985   551.4′57   84-22064

ISBN 0 471 90646 8

British Library Cataloguing in Publication Data:
Bird, Eric C. F.
  Coastline changes : a global review.
  1. Coasts
  I. Title
  551.4′57   GB451.2

ISBN 0 471 90646 8

Phototypeset in 10/12 Linotron Times
by Input Typesetting Ltd, London
Printed by Page Bros. (Norwich) Ltd, Norwich

Contents

Preface ix

Chapter 1 Introduction 1

Chapter 2 Evidence of coastline change 12
    Alaska 12
    British Columbia 18
    USA—Pacific Coast 18
    Pacific Mexico 25
    Pacific Central America 25
    Pacific Colombia 26
    Ecuador 27
    Peru 27
    Chile 27
    Argentina 29
    Uruguay 29
    Brazil 30
    French Guiana, Surinam and Guyana 31
    Venezuela 31
    Caribbean Colombia 31
    Caribbean Central America 32
    Caribbean Mexico 32
    USA—Gulf Coast 33
    Caribbean Island Coasts 36
    USA—Atlantic Coast 37
    Canada—Atlantic Coast 40
    Canada—Arctic Coast 41
    Great Lakes 42
    Greenland 43
    Iceland 44
    Norway 44
    Sweden 46
    Finland 48
    Baltic USSR, Estonia, Latvia and Lithuania 49
    Poland 50

East Germany and Baltic West Germany 50
Denmark 51
West Germany—North Sea Coast 57
Netherlands 57
Belgium 59
British Isles 59
France 76
Spain and Portugal 78
Mediterranean France 79
Corsica 80
Italy 80
Malta 83
Yugoslavia 83
Albania 84
Greece 84
Western Black Sea 87
USSR Black Sea Coast 89
Caspian Coast 91
Turkey and Cyprus 95
Syria and Lebanon 97
Israel 97
Egypt 98
Libya 99
Tunisia 99
Algeria 100
Morocco 100
West Africa 101
South Africa 105
Moçambique 106
Madagascar 107
Tanzania 107
Kenya 107
Somalia and Djibouti 108
Red Sea Coasts 108
Southern Arabia 108
Arabian Gulf Coasts 109
Iran 109
Pakistan 110
India 110
Sri Lanka 111
Bangladesh 112
Burma 112
Western Thailand 113
Malaysia and Singapore 113
Eastern Thailand 116
Kampuchea 116
Vietnam 117
China and Hong Kong 117
Taiwan 119

Korea 119
Japan 119
Pacific USSR 123
Arctic USSR 124
Philippines 124
Indonesia 126
Papua New Guinea 132
Australia 134
New Zealand 147
New Caledonia 151
Fiji 152
Hawaii 152
Tahiti 154
Other Pacific Islands 154
Atlantic Ocean Islands 154
Indian Ocean Islands 156
Antarctica 156

Chapter 3 **Categories of coastal change** 158
Cliffed coastlines 158
Glaciated and periglaciated coastlines 159
Emerging and submerging coastlines 160
Volcanic coastlines 161
Landslides 161
Deltaic coastlines 162
Beaches, spits and barrier coastlines 163
Erosion of sandy coastlines 168
Swampy coastlines 174
Artificial coastlines 175

Bibliography 177

Appendix 193

Author Index 196

Location Index 200

Subject Index 217

Preface

This book is based on the results of a project carried out by the International Geographical Union's Working Group on the Dynamics of Coastline Erosion (1972–76) and its succeeding Commission on the Coastal Environment (1976–84). As Chairman of the Working Group, and subsequently of the Commission, the author acted as Convenor for world-wide studies of coastline changes during the past century, and for longer periods where suitable information could be obtained. The outcome was the collection of a large amount of data on the nature, extent, and history of coastline changes, supplied by over 200 correspondents representing 127 coastal countries. In the course of this work it became clear that erosion has been more extensive than deposition around the world's coastline in recent decades, especially on low-lying sandy coasts. The explanation for this is not simple: a number of factors have contributed to the modern prevalence of erosion, their relative significance having varied from one section of coastline to another. It also became clear that, in our present state of knowledge of the world's coastline (some parts of which have been intensively studied, while others are known only at reconnaissance level), any attempt to derive globally valid generalizations from research on any one section of coastline is hazardous. Our understanding of coastal processes is still based on selective studies of very limited parts of the world's coastline: the opportunities for geomorphological research on the less well known sectors are extensive.

It is hoped that this book will stimulate interest in the study of coastal geomorphology, starting from the recognition, mapping, measurement and analysis of changes in progress on coastlines. It is impossible to report all the evidence of historical changes on the world's coastlines in such a book as this. Instead, the aim is to illustrate the kinds of change that have been documented around the world's coastline, with selected examples, and references to more detailed work. The first chapter reviews the origins of the International Geographical Union's project, examining the problems and methods of documenting geomorphological changes on particular coastlines over selected periods. There follows a round-the-world selective summary of the record of coastline changes, based on published references and unpublished reports by individual members of the Commission on the Coastal Environment, whose names are given in the text, with affiliations listed in the Appendix on page 193. This is illustrated by selected maps and photographs, but readers will need to use a good atlas, or consult national maps of coastal areas to locate places mentioned in this review. Reference can also be made to the illustrated descriptions of coastal geomorphology assembled in the same sequence in *The World's Coastline* (Bird and Schwartz, 1985), as a background to the study of changing coastlines. The third chapter then reviews the categories of coastline change that were identified in the course of the global project, outlining the geomorphological explanations. Only a few of the sectors where coastline changes have occurred during the past century have so far been investigated in detail, and the task for future research is to determine which of these geomorphological explanations are relevant to particular coastlines, and to assess the relative significance of the various geomorphological processes that have resulted in gains or losses of coastal land.

It is thus hoped that this book will stimulate more intensive local research, using historical sources as well as geomorphological methods, to analyse the changes that have occurred on particular coastlines. As well as people working from universities, colleges, research institutes and field studies centres in coastal regions, school teachers and other interested people living on or near the coast can assemble valuable information. The IGU Commission on the Coastal Environment was able to use a few local studies of this kind. For example, the evidence of stages in the growth of the Pantai Laut spit on the shores of the Kelantan delta, north-east Malaysia (shown in Fig. 59) came partly from studies carried out by a local school teacher.

It is also important to record existing coastal features by means of field surveys as well as ground and air photography, as a basis for measuring subsequent changes. Retrievable historical information on coastlines is patchy and of varying reliability, and more accurate contemporary surveys will benefit future studies. Monitoring of coastline changes is necessary for scientific understanding, as a background for assessing human impacts, and as a means of devising management strategies in the future. The remarkable spread of coastal engineering works in recent decades has already made long sectors of coastline artificial. The debate on whether such works are really necessary, and if they are, which of the possible alternatives (such as artificial beach renourishment) are most desirable, is best founded upon a broad understanding of the geomorphology of coastlines, and their recent evolutionary history. This book is intended to provide a global perspective for such discussion.

I would like to acknowledge the support of the members of the Commission on the Coastal Environment, especially those whose contributions are included in the text. In addition, I would like to thank Jock Murphy, Map Curator of the Baillieu Library, University of Melbourne, for much patient help with maps and charts from various parts of the world. Robert Bartlett and Wendy Nicol, of the Department of Geography, University of Melbourne, assisted by drawing the line illustrations and preparing the photographs respectively, and I am grateful to Neville Rosengren, of the same Department, for critically reviewing the text. In selecting material from the vast amount of information assembled by the Commission on the Coastal Environment I had incidental help from Catherine, Philippa and Jennifer Bird.

Lyme Regis, August 1984   ERIC C. F. BIRD

CHAPTER ONE

Introduction

In July 1972 a group of coastal geomorphologists met in Halifax, Nova Scotia, to discuss the progress of research in their subject. They came to the conclusion that, although there had been a great deal of work on changes of sea level, and on upward and downward movements of the land, around the world's coasts relatively little attention had been given to the advance and retreat of coastlines, the gains and losses of land that result either from changes of sea level relative to the land or to erosion and deposition. Dr Hartmut Valentin, who was present at this meeting, had dealt with coastline changes in general and theoretical terms in his treatise, *Die Küsten der Erde*, and had devised the well-known analytical diagram reproduced here (Fig. 1); there had been many detailed studies of coastline changes at particular localities around the world, but no attempt had been made to document, measure, and analyse such changes on a global scale or over a particular interval of time. Accordingly, the Halifax group recommended a project on changes around the world's coastline during the past century, and the ensuing 22nd International Geographical Congress, held in Montreal, set up a Working Group to compile this information. Four years later a preliminary report was prepared (Bird, 1976) and widely circulated at the 23rd International Geographical Congress, in Moscow. As a sequel, the Working Group became the Commission on the Coastal Environment, which has carried out a number of projects, including further documentation of coastline changes. The results of this work have been presented in further reports (Bird, 1980), and are reviewed in this book.

The time scale of a century was originally selected because it was known that maps and charts dating from the period 1870–1900 were available in

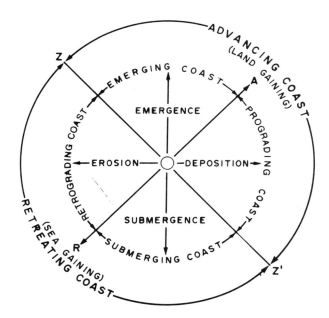

Fig. 1 Analysis of coastline change (after Valentin, 1952). Z–Z′ indicates a stable coastline, where erosion has been offset by emergence, or deposition by submergence, or where no changes have taken place (O). It is necessary to specify a time scale over which changes have taken place before using this scheme, for many coastlines have shown alternations of advance and retreat during the past century

many parts of the world, and that coastline changes could be detected and measured by comparing these with later surveys and modern air photographs. In some countries, such as Britain and Denmark, where earlier maps and charts are available, it was possible to determine changes over a longer period. In the Mediterranean region, and locally elsewhere, there is sporadic evidence of coastline changes over periods of at least 2000 years from historical descriptions and datable archaeological evidence. Some of these ancient sites, particularly in Greece and Turkey, were coastal settlements that are now found some distance inland, as the result of coastal deposition perhaps accompanied by emergence, due to land uplift or a fall in sea level, leading to a seaward advance of the coastline. Elsewhere, the foundations of ancient settlements are out on the sea floor, as the result of submergence, or the cutting back of the coastal margin by erosion. Still longer spans of coastline change can be determined where former shore deposits (beaches, corals, salt marsh) now found inland, perhaps above sea level, or offshore on the sea floor include material (such as shells, wood or peat) that can be dated by radiocarbon or other geochronological techniques. On some coasts it is possible to trace the extent of gains and losses of land since the sea attained approximately its present level (about 6000 years ago) as a sequel to the world-wide Holocene marine transgression, the sea level rise that began about 18,000 years ago, when the cold global climates of the Pleistocene ice age started to become warmer. It is important to realize that the existing world coastline has been shaped largely within this 6000-year period, with the sea at, or close to, its present level in relation to the land. Further details of this coastal evolution may be obtained from geomorphological textbooks (e.g. Bird, 1984).

Although information on coastline changes during the past century is widely available from historical maps, charts and photographs there are still many countries, especially in polar and tropical regions, where reliable surveys of coastlines exist only for the past few decades, usually the period for which air photography is available. In some cases the best available evidence of coastline change is purely geomorphological: beach-ridge plains, deltas and marshlands that have clearly prograded, or cliffs that have been cut back. Alternations of advance and retreat are indicated where coastal plains that had formerly been built forward by deposition now show eroding seaward margins (Fig. 2), or where earlier cliff recession has been brought to a halt by the accumulation of a wide beach, or beach ridges, in front of an abandoned bluff (Fig. 3). Such landform features may indicate the kind of changes that have taken place, but evidence of *rates* of coastline change require studies of maps or charts of various dates, successive ground or air

Fig. 2 Recession of a formerly prograded sandy coastline is indicated by a cliff cut into previously built beach ridges at Sandy Point, Victoria, Australia. Photo: Eric Bird (December 1979)

Fig. 3 Beach progradation has taken place in front of a formerly cliffed, receding coastline at Twilight Cove, Western Australia. Photo: J. N. Jennings (August 1963)

photographs, or even (making due allowance for artistic interpretation) dated drawings or paintings which show the former coastal configuration. In a few cases, written accounts of changing coastlines have proved useful: for example historical records of the lighthouse at Cap d'Ailly in northern France, which in 1775 was built 160 metres inland behind a retreating cliff. In 1845 it was only 60 metres inland, and in 1940, when it was destroyed by bombing, it stood on the cliff edge. From this, Ottmann (1965) deduced that cliff recession had averaged about a metre per year. In North Carolina the site of Sir Walter Raleigh's English colony (1585–7) on Roanoke Island is well documented, but failure to find it has been attributed to cliff recession, which measured 282 metres between 1851 and 1970, and could have attained 600 metres since Raleigh's time, thereby destroying this colonial site (Dolan and Bosserman, 1972).

In such studies it has been necessary to define the coastline on which changes have been measured. The term *coastline* is here taken as the seaward margin of the land, whereas the term *shoreline* denotes the water's edge, which moves to and fro as the tides rise and fall. The coastline, thus defined, is usually equivalent to the high spring tide shoreline, but where the tide range is large, as in the Bristol Channel or the Bay of Fundy, there is considerable variation in the positions reached by the sea at spring tides. Moreover, meteorological effects, including storm surges, and other unusual events such as tsunami waves generated by volcanic eruptions or earthquakes, result in temporary submergence of coastal land margins, and sometimes achieve considerable geomorphological change along the coastline. In practice, measurements of change have usually been made with reference to the crest or base of a cliff, or the seaward limit of backshore vegetation on beaches and deltas. Where the shore is occupied by salt marshes or mangrove swamps the vegetation boundary is usually well-defined, especially where erosion has cut a small cliff along the seaward margin. Such features are found at various inter-tidal levels, and changes mapped along them are strictly shoreline rather than coastline changes. Within the inter-tidal zone it is possible for high-tide shorelines to advance at the same time as low-tide shorelines retreat, the shore profile becoming steeper, and *vice-versa* as the shore profile is flattened (Fig. 4). Given these complications, changes are best mapped and measured along *coastlines*, as defined here.

Coastline changes can be expressed in linear terms, as advance or retreat measured at right-angles to the coastline; in areal terms, as the extent of land gained or lost from a coastal sector; or in volumetric terms, as the quantity of material added to, or lost from, the coast. Most reports of coastline advance or retreat have been based on linear or areal measurements, but reference will be made to attempts to assess 'sediment budgets' in terms of volumes of material eroded, transported and deposited within a coastal environment: there have actually been very few of these (cf. Figs 39, 43). Volumetric studies are easy to advocate, but in practice the difficulty of mapping and measuring changes in the nearshore zone, and out on the sea floor, has hampered attempts to quantify sediment movement, and

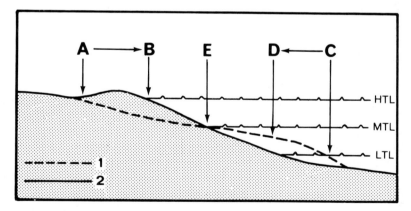

Fig. 4 Shorelines can be defined as the water's edge at various stages of the tide (e.g. high tide level, mid-tide level, and low tide level, as shown here). They may all advance seaward on a prograding coast, or retreat landward on a retrograding coast, but independent migrations can occur. Thus an advance of the coastline (high tide shoreline) from A to B may be accompanied by a retreat of the low tide shoreline from C to D, with no change at mid-tide level. In order to avoid such difficulties, coastline changes are measured as advance or retreat of the high tide shoreline, unless changes within the inter-tidal zone (e.g. on salt marshes) are specifically considered

assessments which treat offshore changes as a 'balancing item' achieve only a partial understanding of the coastal system. In general, coastline changes have been expressed in terms of linear or areal measurements, usually given over specified periods (e.g. 1925–78) determined by the dates of the maps, charts or photographs used, and sometimes expressed as annual averages (e.g. 2.5 metres per year). Annual averages can indicate only the mean trend, for rates of coastline change are often highly variable, with most variation occurring in stormy periods or episodes of coastal flooding by rivers or by the sea. Repeated surveys of low-lying coasts have often shown alternations of advance and retreat, with a resulting gain or loss, or perhaps no net change, over periods of a decade or more.

The problems encountered in using historical maps and charts as evidence of past coastline configuration have been summarized by Johnson (1925) and Carr (1962). Firstly, there is the question of technical accuracy at the time of the original survey, depending on the methods and instruments employed. Maps made before 1750 are generally unreliable, variable in scale and incorrect in detail, but after the introduction of triangulation by theodolite in the late eighteenth century accuracy improved. Nevertheless, errors persisted. Carr quoted an example from the first edition of the British Ordnance Survey map (1809) of part of North Devon which showed a rocky cliff a quarter of a mile *landward* of its 1960 position, commenting that a change of this kind would possibly have been accepted as evidence of progradation in the intervening period had this been on a low-lying sandy coastline. There are also mistakes made by cartographers when the map was being drafted, especially where accurate portrayal of the coastline was not essential for the purposes of the original map.

Secondly there is the problem of partial revision, when a new edition of a map retains some unrevised outlines from its predecessor. Coastal configuration may thus be shown as unaltered when in fact a change had occurred by the time the new edition was produced. In some cases the date on a map is the date of drafting or publication, rather than the survey on which the map was based.

Hydrographic charts are intended to show the pattern of navigable waters and the locations of hazards to shipping, and they are often inaccurate or out of date in portraying the coastline, especially where it is low-lying and the nearshore zone broad and shallow. Coastlines are usually more accurate on land maps, but these may be unreliable in showing inter-tidal features, especially low-tide shorelines, which are generally submerged and less easily surveyed and revised than features on land. The inter-tidal zone is still in many respects a no-man's-land where there is scope for the improvement of techniques of field surveying, remote sensing, and cartographic representation: the evidence for past positions of inter-tidal shorelines is much less reliable than the relatively easily mapped land margin, the coastline.

Finally there are the changes that occur after the publication of a map or chart, notably the stretching, shrinkage or distortion of the paper on which it is printed or on to which it has been copied by photographic or other methods. These effects can be minimized by the use of more stable materials, and by storage in dry, constant-temperature environments, but while this may provide more accurate data in the future it cannot make good the depredations of the past. In some cases it may be possible to retrieve the accuracy of a distorted historical map, if the original survey data can be found and used in re-drafting.

In recent decades, mapping has been based increasingly on air photography, using photogrammetric methods (with field checking of the nature, location, altitude and spacing of ground features) to produce original maps and charts, or to revise earlier surveys. Modern maps are usually more accurate than their predecessors, produced entirely by field surveys, and when they are used in the study of coastline changes it should be possible to refer back to the original air photography to check details and verify measurements (Stafford, 1972; El Ashry, 1977).

Measurements made from air photographs must also take account of internal scale variations due to (1) radial distortion towards the margins of the photograph, so that measurements should preferably be made from photographs where the features concerned are centrally placed; (2) relief distortion due to the portrayal of a variable surface topography on a flat plane, which is more of a problem on steep or cliffed coasts than on low-lying beach-fringed, swampy or deltaic coasts; and (3) tilt distortion where the airborne camera was not strictly vertical when the photography was taken, or where scale variations in a run of air photographs result from an ascent or descent of the aircraft. As with maps, there are errors introduced by stretching, shrinkage or buckling of the film or the printed photography.

In some cases it has been possible to make corrections for these various errors by using stereographic plotting instruments that can readjust distorted imagery with reference to the spatial distribution of ground control points. One must be sure that these points remained unchanged over the period of study: variations due to road widening, fence realignment, modifications of buildings and growth of tree canopies can lead to error when coastline changes are measured with reference to them (Fisher and Regan, 1978).

Despite enthusiastic advocacy of the value of satellite imagery in coastal studies on the part of agencies producing this material, it has so far proved to be of limited use in measuring coastline changes. The dimensions of pixels, unit areas of remote sensing, determine the precision with which a coastline can be located, for irrespective of the position of the coastline a pixel records land if 50% or more of its area of land. On the one hand a slight advance of a coastline may increase the land area within a pixel from 49% (recorded as sea) to 51% (recorded as land), the satellite imagery registering an apparent advance by the width or breadth of a pixel. On the other hand a substantial advance of a coastline, from halfway across one pixel to almost halfway across the next, will fail to register on satellite imagery. As Landsat imagery used pixels of approximately 60 metres by 80 metres, successive imagery cannot demonstrate changes within ±30 metres to ±40 metres, so that only the most rapid and extensive changes can be detected. Satellite photography has proved useful in estimating *areas* of land gained or lost, for example on the deltaic islands of the Bangladesh coast (Polcyn, 1981), and changes in the area covered by glaciers in Alaska and on the margins of Antarctica, but it is necessary to use conventional air photography to map the actual advance or retreat of coastlines accompanying such gains or losses. Undoubtedly techniques of mapping linear features from satellite imagery will improve, but in the meantime conventional air photography has been of much more value in detecting and measuring coastline changes than remote sensing from satellites.

DOCUMENTED COASTLINE CHANGES

A great deal of information on the advance and retreat of the world's coastline exists in the form of widely scattered published and unpublished material, much of the latter being held in the archives of national, provincial and local government departments, particularly land survey, port authority and coastal engineering divisions. Geologists, geomorphologists and engineers have long been aware of coastal changes in progress, but systematic studies have been sporadic. In the classic works of Gulliver (1899) and Johnson (1919) there are few references to actual measurements of coastline advance or retreat, and most subsequent textbooks of coastal geomorphology have given greater emphasis to general and theoretical modes of coastal evolution than to the documentation and analysis of actual changes.

One of the most comprehensive accounts of coastline changes was that assembled by the Royal Commission on Coast Erosion in Britain, which drew upon the numerous local reports of the extent and rate of erosion and accretion on the British coastline in the nineteenth century to produce two volumes of evidence (1907, 1909) and a report presenting conclusions (1911). The Royal Commission had set out 'to reach some conclusions with regard to the amount of land which has been lost in recent years by the encroachment of the sea on the coasts of the United Kingdom* and to the amount which had been gained by reclamation or accretion from the sea'. Evidence of changes during the previous century was sought from comparisons of Ordnance Survey maps of various dates, chiefly on the scale of 6 inches to the mile (1 : 10,560) and 25 inches to the mile (1 : 2534), and from information provided by local authorities and private individuals, notably coastal landowners. This was an unprecedented attempt to measure coastline changes on a national scale, and it still remains the most comprehensive study on such a scale in the coastal literature.

The Ordnance Survey provided data on areas of land gained or lost, based on measurements of changes in the position of the high tide line shown on early nineteenth-century surveys and on the most recent revisions then

* The United Kingdom then included all of England, Wales, Scotland and Ireland.

available, dating from the 1890s and the early 1900s. One difficulty was that until 1868 in England, and 1889 in Ireland, the surveyors had mapped the upper and lower limits of ordinary spring tides, but subsequently they used high and low mean tide lines in coastal survey work. Comparisons of maps showing the high spring tide shoreline with those showing the mean high tide shoreline led to over-estimates of land gains and under-estimates of land losses, and in places minor gains were recorded where in fact there had been no change, or even a slight land loss.

Despite these problems the results were considered worth tabulating on a county basis. They showed that Yorkshire, for example, lost 774 acres (313 hectares) between 1848 and 1893, much of it from the cliffed coastline between Bridlington and Spurn Head (identified as one of the more rapidly eroding sectors in Britain), but in the same period gained 2178 acres (881 hectares) by accretion and reclamation, mostly around the Humber estuary. Suffolk lost 518 acres (209 hectares) by erosion and gained only 151 acres (61 hectares) by accretion and reclamation during the period 1879 to 1904, but in general the land that had been gained in estuarine areas exceeded that lost, mainly from coastlines exposed to the open sea. Totalled by countries, the results are shown in Table 1, and showed a substantial excess of land gained by accretion and reclamation over losses of erosion, but if attention had been confined to 'outer coastlines', directly exposed to wave attack from the Atlantic Ocean, the North and Irish Seas, and the English Channel, land lost by erosion greatly exceeded land gained by deposition and reclamation, despite the extensive construction of sea walls and groynes to counter erosion (Sherlock, 1922). The discrepancy is because there had been such extensive land gains by siltation and reclamation within inlets and estuaries, especially around the Wash.

	Land gained (ha)	Land lost (ha)	Net change (ha)
England and Wales	14,344	1,899	+12,445
Scotland	1,904	330	+1,576
Ireland	3,178	458	+2,719
Totals:	19,426	2,687	+16,738

Source: Data compiled by the Royal Commission on Coastal Erosion (1911) for a period averaging 35 years in the nineteenth century.

The Royal Commission also received from the Ordnance Survey measurements of changes in the width of the shore, based on variations in the position of high and low tide lines between early and late nineteenth century surveys. These indicated a reduction in shore area around Britain during that period, implying that the transverse gradient of the inter-tidal zone had steepened. However, changes in the definition of tidal limits used by surveyors, mentioned above, and limitations in the accuracy of field surveys, especially of low tide alignments, make this conclusion doubtful.

The work of the Royal Commission on Coastal Erosion has been followed by many detailed and local studies of historical changes on parts of the British coastline, summarized by Steers (1964, 1973) and by Bird and May (1976). In general there has been a continuation of the trends identified in the nineteenth century, with an extension of artificial coastlines resulting from reclamation and anti-erosion works.

Few coastlines have been as intensively studied as that of the British Isles, but in the Netherlands, Edelman (1977) summarized systematic measurements of coastline advance and retreat over the period 1860 to 1960 (see page 58) and in the United States Shepard and Wanless (1971) reviewed historical evidence of coastline changes over the past two centuries, illustrated with ground and air photographs (vertical and oblique), and a few maps showing historical variations. More recently the 'National Shoreline Study' prepared by the US Army Corps of Engineers in 1968–71 indicated that of the 134,984 kilometres of United States coastline, 32,800 (24%) could be categorized as 'seriously eroding'. Subsequently, May *et al.* (1982) developed a Coastal Erosion Information System for the United States, based on data obtained from geologists and engineers for grid cells of 3 minute latitude and longitude enclosing coastal sectors, of which there are 1689. Although such a system gives only a generalized picture of patterns of coastline change, the prevalence of erosion was clearly demonstrated: on the Atlantic coast it averaged 0.8 metres per year, on the Gulf coast 1.8 metres per year, and on

Fig. 5 Sand deposition occurred in Caraguatatuba Bay, Brazil, after large quantities of sediment were carried down to the sea by flooding rivers following heavy rains in the high hinterland in 1967. Subsequently, wave action has moved sand shoreward to prograde the beaches. Photo: Eric Bird (August 1982)

the Pacific coast (including Alaska) 0.005 metres per year. Such averages, of course, mask a great deal of local and temporal variation, which can only be assessed by means of detailed, local mapping of specific coastal sectors, and the monitoring of continuing changes (Fig. 5). Maps of dated coastline positions have been assembled for only a few sectors of the world's coastline, and standard textbooks of coastal geomorphology have made little use of them, even though analysis of such changes is important, to the under-standing of the evolution and dynamics of coastal systems.

When the IGU Working Group project on coastline changes began in 1972, the first task was to locate and update scattered surveys of changes on particular sectors of the world's coastline. Requests for information on coast-line changes during the past century were sent to correspondents representing coastal countries, and by 1984 over 200 people had supplied information. Much interest was aroused by the accumulated evidence of a modern preva-lence of erosion on the world's sandy coastline, and the attempts to account for this phenomenon (Bird, 1980). However, changes on cliffed and rocky coastlines, on deltas, on volcanic coasts, and on swampy coasts also proved interesting, as did the evidence of rapid artificialization of coastlines, especially in North America, Europe, Japan and Australia. The Working Group, and its successor, the Commission on the Coastal Environment, endeavoured to stimulate and organize research into historical and active changes on coastlines for which little information was available, particularly in Central and South America, tropical Africa, and south-east Asia. As a result, it is now possible to present a broad picture of the pattern of changes on the world's coastlines over recent decades, the kinds of changes that have taken place, and in some cases the rates at which these changes have proceeded. Such a presentation identifies many problems for analysis and explanation by geomorphologists, and the aim of this book is to promote such discussion, providing locations and available references. The following chapter gives information on coastline changes in a round-the-world sequence, the same sequence as is used in *The World's Coastline* edited by E. C. F. Bird and M. L. Schwartz: Van Nostrand Reinhold, 1985), from which background material on national and regional coastline features can be obtained.

CHAPTER TWO

Evidence of coastline change

Evidence of changes on the world's coastline is here described in a sequence that begins on the Arctic coast of Alaska and proceeds counter-clockwise around North and South America to Arctic Canada. Greenland and Iceland are then treated, and another counter-clockwise sequence begins in northern Norway, and proceeds by way of Europe and the Mediterranean, around Africa to India and south-east Asia, China, Japan, and so to the Pacific and Arctic USSR. The Philippines, Papua New Guinea, Australia, New Zealand, and the Pacific Islands are then dealt with, followed by some reference to the smaller Atlantic and Indian Ocean islands, and finally to the coastline of Antarctica. Interruptions are made at appropriate points to include the Great Lakes and the Caspian Sea, and offshore islands such as Britain and Madagascar.

It is of course impossible to deal with all of the evidence of changes on the world's coastline, and a selection has been made of features that are illustrative of the various categories of coastline change that have been identified and documented by the IGU Commission on the Coastal Environment. Chapter Three then analyses the various categories of coastline change, and indicates possible explanations.

ALASKA

The Arctic coast of Alaska has been studied in recent years by teams from the Coastal Studies Institute, Louisiana State University, but their research output has been concerned mainly with present-day beach dynamics and nearshore processes. Historical information is meagre in this part of the world and evidence of coastline changes is based largely on air photographs taken since 1950, and on inferences from existing landforms (Walker, 1985). Short (1979) gave an account of processes at work on the Arctic coastline, where the sea is frozen for several months each year, and waves reach the shore only during the brief summer thaw. The average length of the ice-free season ranges from 4 weeks along the north-facing Beaufort Sea coast to 10 weeks at Cape Lisburne on the west coast. During this period, the waves attack tundra bluffs and rocky outcrops, cutting cliffs, which at Cape Lisburne are up to 275 metres high. Disintegration of tundra bluffs is accelerated by the annual thawing of permafrost, resulting in slumping of material, which is then gradually dispersed by wave action. Lewellen (1970) measured recession rates of up to 10 metres per summer on tundra bluffs south of Elson Lagoon on the Beaufort Sea coast: between 1949 and 1968 a bluff on Flaxman Island retreated 99 metres.

The extensive barrier islands of sand and gravel in northern and western Alaska appear to have been relatively stable in outline over the past few decades except for variations in the position, form and dimensions of intervening inlets. Short (1979) published a series of maps showing changes in the configuration of these barrier islands between 1908 and 1972, due to the breaching, migration and closure of inlets, lateral movement of beach materials, spit growth and truncation, and onshore migration of bars. Longshore drift has carried sediment from eroding to depositional sectors. Depositional capes have been formed, notably at Point Barrow (Fig. 6), where measurements of coastline changes have been made with reference to bench-

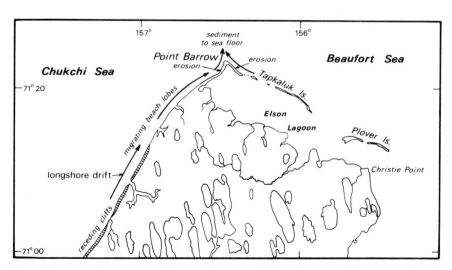

Fig. 6 Recent changes on the coastline near Point Barrow, Alaska. Sand and gravel eroded from the cliffs to the south-west, and from barrier islands to the south-east, is drifting towards Point Barrow, thence out on to the sea floor (after MacCarthy 1953)

marks inserted by the US Coast and Geodetic Survey in 1945, and to land-scape features such as the outlines of lakes and swamps (Rex, 1964; MacCarthy, 1953). Recession of Skull Cliff, a slumping tundra bluff to the south-west of Point Barrow, exposed to summer waves from Chukchi Sea, has been of the order of a metre per year, and north of this there have been alternations of advance (averaging 8.2 metres, 1945–49) and retreat (averaging 5.1 metres, 1949–51). These alternations are due to the formation and atmosphere migration of beach lobes. Beach material drifts towards Cape Barrow from both the Chukchi Sea and Beaufort Sea coasts, but the cusp has not recently prograded. Between 1945 and 1951 both its margins receded by up to 5 metres per year as the result of erosion of beach sediment and its removal to the adjacent sea floor.

The other cuspate points on the north-west coast of Alaska show similar features. Point Franklin is an elongated spit that has grown eastwards as beach material drifts alongshore from receding tunda bluffs at Wainwright, while Point Hope is a sharp cuspate foreland which in recent years has shown progradation along its southern flank.

Rivers have also delivered large quantities of sediment (mainly fine sand and silt) to the north coast of Alaska, the Colville delta showing a crenulate outline produced by deposition of fluvial material at distributary mouths (Walker and McCloy, 1968).

Shepard and Wanless (1971) summarized and illustrated the available evidence of coastline changes on the western and southern coasts of Alaska, where there are several stretches of receding cliff, notably at Cape Thompson. Beaches and barriers are extensive on the west coast, but again historical progradation has been limited to cuspate forelands, such as Cape Krusenstern and Cape Prince of Wales. Near Cape Espenberg a chain of barrier islands present in 1947 had been eroded away by 1962, and there has been re-shaping of the spit at Point Spencer in recent years. Deltas have grown at the mouths of rivers draining into Kotzebue Sound, but comparison of charts made in 1898 with the modern outline shows little change on the coastline of the large delta built by the Yukon and Kuskokwin Rivers, apart from the growth of some alluvial islands near distributary mouths.

Vulcanicity has contributed to the shaping of coastal landforms on parts of the Alaska Peninsula and in the Aleutian Islands, where cliffs are being cut in lava and ash deposits produced by successive eruptions, and some volcanic islands (e.g. Bogoslof Island), have disappeared completely within historic times. Earthquakes have also led to vertical displacement of coast-lines: in the 1964 earthquake, land uplift resulted in coastal emergence, as on the southern shores of Montague Island, uplifted more than 9 metres, and land subsidence in coastal submergence, as on the Kenai Peninsula, and on Homer Spit, in Kachemak Bay (Fig. 7), which sank nearly 2 metres. At Anchorage this earthquake produced a subsidence averaging 2.5 metres, and also caused landslides at Turnagain Heights, where mass movement into the sea locally and temporarily prograded the coastline.

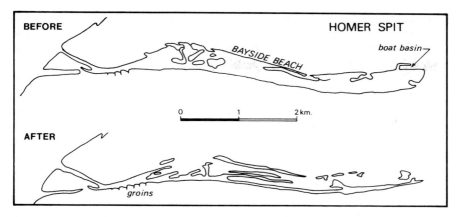

Fig. 7 The coastline of Homer Spit, in Kachemak Bay, Alaska, retreated sharply as the result of submergence when the land subsided during the 1964 earthquake (after Walker 1985)

In Cook Inlet the exposure of high sand bluffs to direct wave attack at high tide has resulted in rapid erosion (US Corps of Engineers, 1971). There have also been rapid changes along the Gulf of Alaska coast especially in and around inlets and embayments, since maps and charts were compiled at the end of the eighteenth and beginning of the nineteenth century. These included charts made by the Russian American Company, 39 of which were incorporated in the Tebenkov Atlas in 1852. Molnia (1979) compared the outlines shown on these with modern cartography and air photographs. Rapid sedimentation, largely outwash from melting glaciers, is filling several inlets, notably Controller Bay, Icy Bay, Yakutat Bay and Lituya Bay, and prograding their coastlines. Where glaciers reach the sea there are stretches of ice coast, many of which have retreated during the past century, their rate of melting having exceeded their rate of advance. In 1904 a large lobe of the Guyot Glacier occupied Icy Bay, and protruded about 5 kilometres into the Gulf of Alaska. Subsequent recession of the ice front (Fig. 8) has opened up a branching fiord about 40 kilometres long (Molnia, 1977). Point Riou is a spit that formed at its mouth in 1904 and had grown to about 8 kilometres by 1975 as the result of wave deposition of morainic material supplied from the melting Malaspina Glacier, which reaches the coastline to the east (Nummedal and Stephen, 1976). By contrast, the coastal plain west of Icy Bay is no longer receiving such sediment from the east, and is eroding. Similar features are seen in Yakutat Bay, below the snouts of the Hubbard and Turner glaciers, and in Lituya Bay, while in Taylor Bay the deposition of an outwash plain in front of the Brady Glacier (Fig. 9) prograded the coastline by 4 kilometres between 1926 and 1977 (Molnia, 1979). In 1958 an earthquake caused a massive landslide that dumped more than 30 million cubic metres of rock debris into Lituya Bay, producing giant waves that

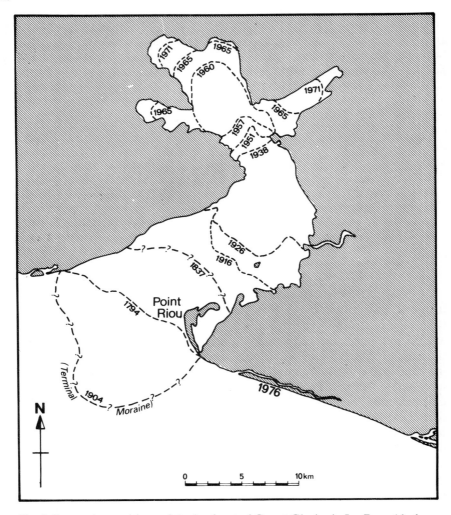

Fig. 8 Successive positions of the ice front of Guyot Glacier in Icy Bay, Alaska, in relation to the 1976 coastline (Molnia, 1977). The 1794 limit is taken from Vancouver's chart, and the glacier then retreated to approximately the 1837 alignment before a late nineteenth-century readvance formed a protrusion marked by the 1904 terminal moraine. Intermittent retreat of the ice front subsequently has opened up Icy Bay as a fiord, and by 1976 longshore drifting of beach material from the east had shaped a substantial spit at Point Riou

swept down the fiord and out over La Chausée spit at its entrance. As in previous such events (1853–4, 1874, 1899, 1936) the waves trimmed away the forest on bordering slopes, but there was little change around rocky and bouldery shores of Lituya Bay (Miller, 1960).

To the south-east, in Glacier Bay, the Grand Pacific Glacier receded 25 kilometres and the Johns Hopkins Glacier about 16 kilometres between 1890 and 1960, to terminate in ice fronts on the shore at the heads of fiords. Sediment relinquished by the receding ice fronts to glacifluvial streams has prograded deltas, as near Juneau, where the Norris Glacier retreated almost a kilometre between 1890 and 1960, and its outwash has built a delta extending 3 kilometres in front of the ice margin. Sediment carried downstream from glaciers that terminate farther inland has contributed to the continuing growth of deltas such as that of the Susitna at the head of Cook Inlet. In Prince William Sound the Valdez glacier had built a delta, the seaward margin of which was cut back as the result of submarine slope failure during the 1964 earthquake; the Copper River delta to the east was uplifted during this earthquake, so that its seaward margin advanced, and its distributaries became incised into the delta plain.

In south-east Alaska coastal progradation has been assisted by tectonic uplift of the land margin due in part to isostatic recovery of the area of deglaciation proceeds (Hicks and Shofnos, 1965).

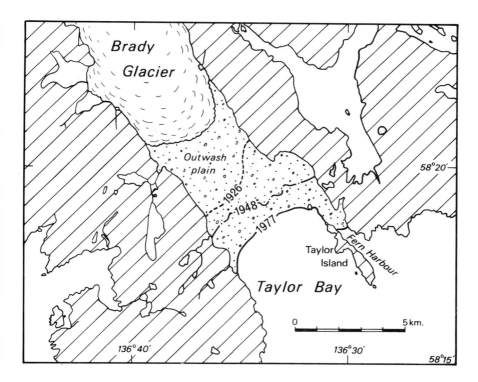

Fig. 9 Map of Taylor Bay showing the progradation of the coastline between 1926 and 1977 as a result of deposition of glacial sand and gravel outwash from the melting Brady Glacier. The volume of deposition has been sufficient to fill water up to 20 metres deep (after Molnia, 1979)

BRITISH COLUMBIA

It is unlikely that much of the steep rocky coastline of British Columbia has changed measurably during the past century, but there have been minor changes on the small river deltas at the heads of fiords, for example the Skeena River delta, growing into an estuarine inlet near Prince Rupert. Along the shores of the Strait of Georgia cliffs are receding in glacial drift deposits, and beaches have been augmented by the arrival of sand and gravel derived from these cliffs, particularly on the east coast of Vancouver Island (McCann and Hale, 1980), and on the spits and cuspate forelands which are numerous around the islands in the southern part of this strait.

The Argonaut Plain, on north-east Graham Island, is an area where progradation is continuing, as sediment derived from the erosion of unconsolidated rock outcrops and glacial drift deposits is carried alongshore and built into successive beach ridges, converging in the cuspate foreland at Rose Point (Owens and Harper, 1985).

The Fraser delta is also prograding as salt marshes spread forward on to depositional inter-tidal mudflats, which are built up episodically by sedimentation from river floodwaters (Luternauer, 1980). Changes in the deltaic outline (1827, 1860, 1890, 1919) mapped by Johnston (1921) included the growth of Westham Island, and between 1929 and 1959 the deltaic coastline advanced up to 2.3 metres per year.

The ocean coast of Vancouver Island is mainly steep and rocky, intersected by fiords, but with some beach-fringed sectors. Little change has occurred here within historical times.

USA—PACIFIC COAST

South of the Canadian border at Boundary Bay the coastal landscape is dominated by glacial drift deposits. Cliffs cut into these deposits on promontories and islands have yielded sand and gravel deposits, which have been carried alongshore and built into beaches and spits, several of which are still growing. At Shannon Point, on Fidalgo Island, the cliff retreated about 16 metres between 1893 and 1969 (Schwartz, 1971). There has been growth of deltas in the relatively sheltered waters of the southern part of the Strait of Georgia, for example on the Nooksack River delta, which advanced its depositional coastline into Bellingham Bay by almost a kilometre between 1906 and 1961 (Shepard and Wanless, 1971). The nearby Lummi River delta ceased to grow after its outflow was diverted into the Nooksack River by a log jam in about 1860. Cliffs, bluffs, spits and small deltas are seen farther south, in Puget Sound, while on the southern shores of the Strait of San Juan de Fuca, Dungeness Spit is a complex depositional formation, with changes continuing along its outer coastline and at the end of its several hooks. To the west, Ediz Hook was a similar but simpler spit, formed of sand and gravel carried eastwards from the Elwha River and eroded from cliffs of glacial drift. The building of a dam on the Elwha River, and a water supply pipe in front of the cliffs, in the 1930s diminished the sediment supply and resulted in severe erosion of the spit. From 1937 onwards, progressive boulder armouring, groyne construction and shore walling has been necessary to hold the coastline, and in 1975 an artificial beach was added. Deprived of its sediment source, Ediz Hook can only be maintained by such engineering works: it is now essentially an artificial breakwater sheltering Port Angeles harbour.

At and south from Cape Flattery the Pacific coast consists of intermittently receding cliffs and bay beaches with sandy material derived from nearby cliff erosion. Southwards the beaches lengthen, supplied with sediment from slumping Tertiary sandstones and glacial drift deposits. At Cape Shoalwater, alongside the entrance to Willapa Bay, the cliffs have receded up to 3.2 metres in 90 years as a result of undercutting by waves, the nearshore water being deepened by tidal scour as the entrance channel migrated towards the north shore. Beaches to the south have prograded with the arrival of sandy sediment drifting up the coast from the mouth of the Columbia River, at the Oregon border (Schwartz and Terich, 1985).

On the Oregon coast beaches in receipt of sediment from the Columbia River have prograded. In detail, the pattern of deposition around the mouth of the Columbia has been modified by the construction of jetties bordering the river mouth between 1885 and 1913. A new beach (Peacock Spit) has formed between the northern jetty and Cape Disappointment, while to the south of the southern jetty Clatsop Spit has prograded by up to 3 kilometres on a sector about 50 kilometres long, some 15 beach and dune ridges marking stages in this man-induced coastline progradation (Russell, 1970). Excessive sand accumulation at Seaside, a resort at the southern end of this beach, has necessitated regular removal of sand from the promenade. Residents have complained of the lengthening walk to the surf because of the 150 metres progradation between 1944 and 1960, as well as of seaview losses because of increasing dune heights (Stembridge, 1976). On the global scale, these are most unusual problems for a seaside resort: the more usual story is of battles against cliff recession and beach depletion. Since 1961 there has been erosion on Clatsop Spit, parts of the formerly prograded coastline being cut back up to 250 metres by 1968. Stembridge (1976) surmised that this was the beginning of erosion due to the entrapment of sandy sediment behind dams in the Columbia River basin, and a consequent diminution of the river-mouth sand supply.

Erosion on the spit at the mouth of the Siletz River may have been accentuated by beach sand mining, while the Nestucca spit to the north was overwashed and breached in a 1978 storm (Komar, 1985). There are several examples of local accretion of beach material alongside breakwaters built at harbour entrances. At Yaquina Bay, northward-drifting sand has been intercepted by a jetty built in 1897 and subsequently extended, and erosion has ensued at Newport, further north. On the other hand accretion north of the breakwater at Tillamook Bay, built in 1969–74, has been followed by erosion on the Bayocean Spit to the south, and predominant southward drifting is indicated by the growth of the Alsea sand spit, widened and prolonged by up to 150 metres between 1969 and 1974 (Stembridge, 1975). Here, as elsewhere on the Oregon coastline, driftwood accumulation has aided sand accretion. Farther south, the northward drifting of beach material is a summer phenomenon, and there is southward drifting in winter. The pattern of accretion alongside intercepting structures changes in consequence: at Bandon there has been beach progradation on both sides of the jetties built at the mouth of the Coquille River, and the pattern is similar at Rogue River and Suislaw River. The coastline advanced at least 60 metres between 1939 and 1967 just north of the Rogue River mouth, forming a sandy plain 150 metres wide in front of bluffs that were eroding sea cliffs early in this century (Stembridge, 1976).

Recession of cliffed coasts in soft formations is in evidence at several places. At Tillamook Head the coastline has alternated as the result of recurrent landslides which formed lobes that were then cut away by marine erosion. Near Cannon Beach, Shepard and Wanless (1971) found that a cliff had been cut back up to 10 metres between 1950 and 1968, while at Brookings the cliff receded 6 metres in 38 years. Near Newport a headland of Miocene sandstone has been isolated as a stack and dissected into a natural arch in stages recorded photographically by Byrne (1964). Stembridge (1976) mapped coastline recession averaging 30 metres, and locally more than 150 metres, at Newport between 1868 and 1967, while to the south cliff recession averaging 1.3 metres per year has undermined the coast road near Yachats. Shale and sandstone cliffs are receding near Cape Blanco, and at Humbug Mountain a major rock slide has formed a persistent promontory.

About two-thirds of the 2900 kilometres California coastline is cliffed or rocky, the remainder being sandy or marshy. Orme (1985) calculated that 85% of the coastline was actively eroding, and anti-erosion works have been introduced on many sectors. Artificial structures are now extensive along urban waterfronts.

In northernmost California, Cape Mendocino is a sector of steep coast, and there are slowly receding cliffs and outlying stacks south to Punta Gorda, and on to Point Delgado. At the mouth of Big Flat Creek the cliffed coast is interrupted by a deltaic fan that has been built seaward about 500 metres. The steep, mountainous coast continues to Cape Vizcaino, but little is known of recession rates here. J. W. Johnson compared photographs of cliffed and rocky shores near Fort Bragg and Mendocino taken in 1860 with the features seen in 1960, demonstrating the erosion and dissection of stacks and headlands, but on the more resistant strongly folded Miocene rocks of Point Arena, to the south, there has been little change in recent decades.

Shepard and Wanless (1971) illustrated Daetwyler's (1965) work on the growth of Sand Point, on the eastern shore of Tomales Bay. The spit that existed here in 1860 had become enlarged into a cuspate foreland by 1957, as the result of accretion of sand derived from cliffs to the north. Recession has been locally rapid on cliffs cut in soft sedimentary rocks on the northern part of the Point Reyes peninsula, and on part of the curved southern coastline facing Drakes Bay where the central sector is protected by the Limantour sand spit. Erosion has been comparatively slow on the hard rocky shores of the terminal ridge at Point Reyes, but Duxbury Point has retreated 50 metres and Balinas Point 60 metres since the US Coast Survey of 1859.

Even under natural conditions, San Francisco Bay would have been reduced in area as the result of depositional progradation around the mouths of the Sacramento and San Joaquin Rivers. In fact it has been reduced from 1800 square kilometres to 1100 square kilometres in the past two centuries, largely as the result of the deposition of in-washed sediment derived from the goldfields of the Sierra Nevada foothills, especially as a result of nineteenth-century sluicing in the headwaters of the Sacramento River (Gilbert, 1917). There has also been extensive artificial reclamation of bordering salt marshes and shallows.

Recession of cliffs cut in Tertiary sands and clays south of San Francisco has disrupted a former coastal highway, and between Point San Pedro and Point Montara the steep coastal slopes in weathered granite topography, rising over 300 metres above sea level, have shown recurrent slope failure, notably at the Devils Slide, where the coast road has been repeatedly carried away. The bold sandstone headland at Pillar Point marks the beginning of Half Moon Bay, where a long curved beach was relatively stable (allowing for seasonal drift southward in winter and northward in summer) until 1959, when the insertion of breakwaters to enclose a boat harbour concentrated and reflected wave action in such a way as to deplete the beach and initiate coastal erosion on the sector to the south. A cliffed coast from the southern end of Half Moon Bay past Point Año Nuevo to Santa Cruz shows locally rapid erosion and slumping of Tertiary shales and sandstones with cliff recession of up to 0.6 metres per year (Griggs and Johnson, 1979).

Monterey Bay is beach-fringed, the coarse sand having been supplied mainly from the Salinas and Pajaro Rivers, and dispersed along the shore, mainly southward, by waves and currents. There has been little progradation, however, because of the landward loss of sand to form dunes and the seaward loss of sand and gravel into the heads of submarine canyons at Elkhorn Slough, off Monterey, and in Carmel Bay. Changes have been very slow on the promontories of granite which flank Carmel Bay, and culminate in Point Lobos. Point Sur is a rocky headland linked to the mainland by an isthmus bearing sand dunes, where sand drifting from the north has been intercepted to prograde the beach. To the south, the steep Big Sur coast has recurrent landslides, especially on outcrops of shales and greywackes.

South of Ragged Point the coast becomes low and sandy, with extensive dunes, the sand coming mainly from the Arroyo Grande, and from Santa Maria and Santa Ynez rivers. Cliff erosion has occurred at Cayucos, Avila Beach and Shell Beach, and Morro Bay has a lagoon almost enclosed behind a dune-capped barrier spit, with a tidal entrance established by breakwaters. Sand drifting from the north accumulated alongside the breakwater, built in 1930, in such a way as to prograde the beach and form a tombolo attaching Morro Rock to the mainland.

Near Point San Luis the cliffed coast has been intricately dissected, but comparison of photographs taken in 1898 and 1945 showed little change. Pismo Beach and its continuations southward to Point Sal are fed by sandy sediment from rivers, and show landward losses of wind-blown sand. In the absence of artificial structures these beach systems appear to be in equilibrium, with a balance between sand supply and removal. Between Point Sal and Point Conception the coast is again cliffed, with rocky shores and numerous stacks, but the beaches are narrow, and have shown little change in recent decades.

At Point Conception the coast trends eastward along the steep flank of the Santa Ynez Mountains. Miocene sandstones and shales predominate, and locally there have been changes identifiable on dated photographs, for example on the cliffs of siltstone at Refugio State Park, which retreated about 3 metres between 1950 and 1968 (Shepard and Wanless, 1971). During floods the short, steep streams have delivered cobbles and even boulders to form deltaic fans at their mouths.

At Santa Barbara there are receding cliffs of siltstone and shale: near Goleta, recession of 4.6 metres was measured between 1960 and 1968. The beaches have been supplied with sand and gravel delivered by small streams draining to the coast east from Point Conception, and carried south-eastwards by longshore drifting. Demand for a deep-water harbour led to the construction of an L-shaped breakwater in 1927–9, which resulted in accumulation of beach material to the west and the onset of severe erosion to the east, notably at the seaside resorts of Miramar and Sandyland, where at one point the coastline receded 75 metres in 10 years. To the east there are salients representing three former deltas, but the only large prograding delta now is that of the Ventura River.

Outlying San Miguel Island is notable for the deposition of beach sand on its northern shore, and the derivation from this of extensive mobile dune systems, which are locally spilling over the southern shore in such a way as to prograde the beach despite the action of wave scour. At the eastern end of this island drifting sand has accumulated to form a sharp cuspate spit. San Nicolas Island has a similar spit, the history of which has been traced by Norris (1952): it was a lobate foreland on the north-east shore between 1851 and 1879, but it migrated to become a sharp cusp extending eastwards by 1925, and has shown minor changes in alignment subsequently (Shepard and Wanless, 1971). Santa Catalina Island is notable for the high receding cliffs on its ocean shore.

Between the Santa Clara River mouth and Port Hueneme the coastline advanced up to 150 metres between 1856 and 1938, and beaches farther south were maintained despite losses of sediment into the head of the Hueneme submarine canyon. Channel Islands Harbour was excavated on the seaward fringe of the Oxnard Plain in 1961, and there has since been beach prograd-ation on either side of bordering entrance jetties in the lee of an offshore breakwater. In 1939 breakwaters were built to protect the naval harbour at Port Hueneme, and as a result almost all the sand arriving from the north is now diverted offshore into the submarine canyon. To the south-east, beaches depleted of sand drift have subsequently eroded, and within 10 years the coastline a kilometre east of the breakwaters had receded over 200 metres. The coast near Mugu Lagoon was re-shaped between 1856 and 1933, recession of the western sector (at the head of a submarine canyon) being balanced by an advance farther east, but after the building of the Port Hueneme breakwaters beach erosion became prevalent along the whole of this coast.

South of Point Mugu the coast is steep, with some bold, cliffed sandstone headlands, the promontory at Point Dume having intercepted southward-drifting sediment to form a broad beach. Along the Santa Monica Mountains cliff recession and slope failure have repeatedly disrupted the coastal highway: a major earthslide in 1958 at Pacific Palisades formed a debris fan that protruded from the coastline, and has since been cut back by marine erosion.

At Santa Monica an offshore breakwater built in 1933 has refracted waves in such a way as to shape a wide cuspate spit in its lee, alongside the Santa Monica pier (Fig. 10). Beach material drifts southwards along the Los

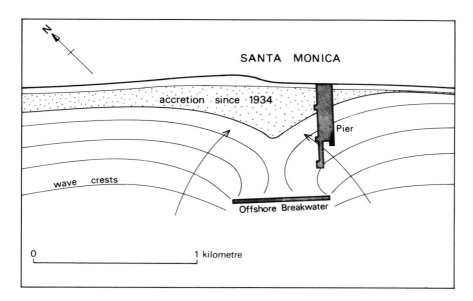

Fig. 10 Localized beach accretion has led to the formation of a cuspate spit in the lee of the offshore breakwater built in 1934 at Santa Monica, California

Angeles coastline, so that construction of a breakwater at Redondo Beach in 1958 led to progradation to the north and loss of beaches and initiation of cliff erosion to the south.

The Palos Verdes Peninsula is bordered by cliffs cut in soft Miocene deposits and harder basalts and sandstones. Landslides occur on the softer outcrops, notably on shales, and on weathered volcanic tuffs near Portuguese Bend, especially during and after rainy weather. In San Pedro Bay, to the south, the San Gabriel River and (after its diversion from a Santa Monica Bay outlet during the 1825 flood) the Los Angeles River have delivered silt and sand to the coast during floods, the sandy fraction being incorporated in the beaches that drift south-east towards Huntingdon.

Beach drifting has been modified following the construction of the break-waters that enclose Long Beach Harbour, and fluvial sediment yields have diminished as a result of the building of reservoirs and storage basins in the river catchments. The Santa Ana River delivers large quantities of sandy sediment during floods, forming shoals off its mouth which are subsequently dispersed by waves and currents, some of the sand moving on to the beaches. Erosion at Newport Beach in recent decades is the outcome of a diminished sediment yield from the Santa Ana River as a sequel to dam building, but minor accretion has occurred both to the north and the south of breakwaters built to stabilize the harbour entrance.

Emery (1960) emphasized the role of dam construction and fluvial sediment interception by reservoirs to explain the prevalence of sandy coastline erosion in Southern California, but Orme (1985) has pointed out the significance of variations in fluvial sediment yield due to short-term climatic fluctuations. Most of the 9.6 million tonnes of sand delivered to the coast by the Santa Clara River between 1933 and 1938 arrived in a 6-day flood in 1938, and in the 1969 floods this river delivered 47.6 million tonnes, compared with 1 to 2 million tonnes in dry years. Beach erosion, prevalent in the dry years 1939–68, gave place to accretion in years when flooding occurs (e.g. 1969, 1978, 1980, 1983), but in recent decades the erosional trend has predominated.

South from Newport there are sectors of cliff cut in shale, sandstone, and conglomerate, the latter being more resistant to erosion. The cliffs are often steep, but rates of recession have been generally slow. Nevertheless, erosion of the sandstones has contributed sediment to adjacent beaches, accompanying the supply of sandy material from rivers. At La Jolla there are well-documented seasonal alternations, the beaches being lowered and cut back by winter storms and built up by summer swell, but the coastline has been maintained over the past few decades.

Cliffed sectors at La Jolla and Point Loma have shown modification during the past century—for example, the natural arch at Cathedral Rock, present in 1900, had disappeared by 1968—but other sectors show little if any change. The curved barrier spit that almost encloses San Diego Bay has been modified following the construction of the harbour jetty in 1900: beach erosion has since been dominant on Coronado Strand, which has been protected by the dumping of boulders and artificially renourished with sandy sediment dredged from the harbour.

PACIFIC MEXICO

South from the Californian border the Mexican ocean coast of Baja California is characterized by alternations of steep and cliffed sectors with beaches and depositional lowlands. Sand accretion continues to prograde the beaches in Sebastian Vizcaino Bay, especially on the southern shores, where some of the sand drifts inland to form barchans in the desert hinterland. More sand accretion is in progress farther south, along the barrier island coast of Isla Magdalena.

On the western shores of the Gulf of California there has been little change on steep coasts in hard igneous rock, but accretion has continued on the cuspate foreland of Punta Arena del Sur. At the head of this Gulf the Colorado River has built a major delta, parts of which are still prograding, and on the low-lying eastern shores sandy deposition has been extensive, especially near the mouths of rivers. The Mayo and the Fuerte Rivers are building large swampy deltas on a coast sheltered from ocean action by Baja California. The Rio Grande de Santiago is also a major sediment-yielding system, and W. F. Tanner* has reported that nearby sectors of beach-fringed coast are advancing, notably in the lee of Tres Marías Islands, and along the seaward margin of the beach-ridge plain fringing the states of Sinaloa and Nayarit. Holocene progradation was extensive between Tepic and Mazatlán, and the beaches at Playa Caimanero are still growing, but accretion appears to be diminishing on this coast, and Tanner (1975) noted evidence of incipient erosion. Out on Isla San Benedicto, Richards (1960) documented the rapid erosion of soft and unconsolidated deposits emplaced on a high wave energy coast by a volcanic eruption in 1952. Initially, cliffs cut in volcanic ash receded by up to a metre per day.

South from Cape Corrientes the coastline is generally steep and rocky, with hard metamorphic formations outcropping, but parallel sandy beach ridges of Holocene age occur along the Acapulco coast, and there are pocket beaches between bedrock headlands near Salina Cruz. These appear to be stable or growing slightly (Tanner, 1975). In contrast with the United States coast, artificial structures are still rare in Mexico, and beach systems remain generally in a natural condition.

PACIFIC CENTRAL AMERICA

In Guatemala the coast consists of sandy beaches and dunes fringing a depositional plain. Minor progradation has occurred around river mouths, but in general erosion has predominated in recent decades. As a sequel to the catastrophic eruption of the Santa Maria volcano in 1902 a vast volume

* Reports quoted without date are unpublished communications to the IGU Commission on the Coastal Environment.

of pyroclastic sediment moved down the Rio Samala valley, and between 1902 and 1922 a deltaic lobe formed, extending 7 kilometres out from the previous coastline. Subsequently this has been cut back by marine erosion, and sandy material has been carried alongshore to prograde adjacent beaches (Paskoff, 1981a).

In 1954–5 Gierloff-Emden carried out detailed surveys of two sectors of the coast of El Salvador, one between the Rio Jiboa and the Rio Grande de San Miguel, and the other of the western shores of the Gulf of Fonseca (Gierloff-Emden, 1959). The first shows a depositional coastline with a seaward fringe of sandy barriers bearing numerous beach ridges in front of extensive estuarine mangrove swamps. River mouths and tidal inlets are migratory, and it appears that sand carried down to the coast by rivers during episodes of flooding is thereafter reworked and incorporated in the barrier beaches. However, the beaches are backed by 'sandsteilkante', steep slopes resulting from marine erosion, and progradation has been confined to spits bordering river mouths and tidal inlets. On the western shores of the Gulf of Fonseca there are receding cliffs cut into the lower slopes of the Conchagua volcano, and within this gulf mangrove-fringed sandflats and mudflats are advancing on the deltaic shores of the Rio Goascoran. Black sands of volcanic origin dominate the beaches here.

The coast of Nicaragua is also bordered by sandy beaches and barriers, on which there has been erosion in recent decades, giving place south-eastwards to steep slopes with rocky shores, along which there have been only minor changes. Similar steep slopes are found along much of the Costa Rican coastline, except for prograding fluvially nourished beaches at the head of the Gulf of Nicoya. Low sandy coasts, interspersed with mangroves, border this gulf, and include the large sand spit, Punta Arenas, now largely built over, its link to the mainland preserved artificially (Battistini and Berg-oeing, 1983). In the Gulf of Panama the coastline is low-lying and swampy, with mangrove progradation around river mouths, and on broad tidal flats on the western shore. A large sand spit is growing at Punta Chame.

PACIFIC COLOMBIA

The northern part of the Pacific coast of Colombia is steep, with minor basal cliffing along a rocky shore, while the southern part consists of mangrove-fringed deltas and estuarine bays (Schwartz, 1985a). Progradation continues on the shores of the San Juan, Patia and Mira deltas. West (1956), describing the mangrove communities noted that their seaward fringe consisted of accreting mudflats and discontinuous sandy beaches. In 1979 an earthquake off Tumaco was accompanied by subsidence of up to 1.6 metres, resulting in coastline retreat by submergence and erosion along a 200 kilometre sector in southern Colombia and adjacent parts of Ecuador (Herd et al., 1981).

ECUADOR

The northern part of the Ecuador coast has extensive mangrove fringes in estuaries and inlets, but in the drier regions south of Bahia de Caráquez beaches are prevalent, interrupted by rocky headlands (Ayon and Jara, 1985). Erosion on the southern part of the beach at Bahia de Caráquez has been accompanied by accretion to the north. At Manta there is recession of cliffs in soft silty clays, and drifting sand has accumulated north and south of the harbour breakwaters. Gradual beach recession has occurred farther to the south, exposing beach rock and undermining the coast road near Salinas. Slumping cliffs border parts of the Gulf of Guayaquil, but deposition of sand and silt from rivers is enlarging deltaic islands, promoting the advance of mangroves, and building up the estuarine shoals south of Guayaquil. The southern shore of the gulf also shows advancing mangroves, and sand spits are growing east from the Tumbes delta.

PERU

In Peru, sandy coasts fringe the Sechura Desert region, and historical progradation of beaches has occurred in the vicinity of the mouth of Rio Tumbes (Broggi, 1946). Much of the coast is steep and cliffed, with beaches derived partly from fluvial sediment delivered to river mouths during occasional floods, and partly from cliffs up to 170 metres high cut in poorly cemented piedmont conglomerates (Dresch, 1961).

Cliff recession is in progress at Cabo Blanco, near Negritos, south of Trujillo, and near Huarmey (Bird and Ramos, 1985). Beaches are generally narrow, consisting of sand and gravel derived from cliffs and wadis, but there are wider accumulations around river mouths, notably on the Rio Camana and Rio Tambo deltas. Sand drifting southwards from Rio Rimac has been deposited alongside the harbour breakwaters at Callao, the port of Lima, and the spit at La Punta is growing out in the lee of Isla San Lorenzo, to the south. There has also been accretion of fluvially supplied sand north of the mouth of the Pisco River, where the beach ridge plain is up to 500 metres wide, and wind-blown sand spilling on to the southern shores of the Bahia de Paracas, to the south, has prograded local beaches (Craig and Psuty, 1968).

CHILE

An analysis of the Chilean coastline has shown that more than 90% is cliffed and rocky, and that changes during the past century have been generally minor, with localised exceptions where cliffs are being cut back in soft rock formations, and where beaches have been built up or cut away (Araya-Vergara, 1982).

In northern Chile there are extensive steep and cliffed coasts, but with the Atacama Desert as hinterland, fluvial sediment yield has been generally meagre. However, the Arica River close to the Peruvian border produced a lobate delta after a rare episode of torrential runoff in 1973, and by 1980 this had been worn back into a cuspate feature (Araya-Vergara, 1985). Cliff recession is in progress, notably at the base of the steep coastal slopes near Iquique, in Tertiary sediments north of Antofagasta, and north and south of Valparaíso.

In central Chile, higher rainfall produces greater runoff, and larger fluvial sediment yields, augmented by deforestation, have supplied more sediment to beaches. Comparison of historical maps with modern air photographs shows progradation of the sandy coast of the Bay of La Ligua between 1835 and 1962, in receipt of fluvial sediment from Rio Petorca and Rio Ligua. At Quintero an island became attached to the coastline by sand deposition between 1875 and 1924, while at Viña del Mar, a ship wrecked in 1871 was behind a prograded beach 96 metres wide by 1903.

At Chañaral the beach prograded 250 metres between 1897 and 1961 as a result of fluvial sediment supply augmented by the waste materials produced from copper mines at Potrerillos and El Salvador, in the headwater regions of the Salado River (Paskoff, 1981a). Sandy coastlines have also prograded by up to 600 metres at San Antonio (1908–32) and by a similar distance at Trinchera-Quivolgo (1876–1944), and there has been sand accretion along-side and within harbours protected by breakwaters at San Antonio and Constitución Cove (Araya-Vergara, 1985). At Coronel the sandy coastline advanced 150 metres between 1854 and 1937, and out on Isla San Maria beach progradation of 300 metres occurred between 1862 and 1920 (Pomar, 1962).

A detailed review by J. F. Araya-Vergara of coastline changes in Central Chile (latitudes 33° to 34° 30'S), based on air photographs taken in 1955, 1963 and 1978, indicated that about half the coastline had remained unchanged, while the other half had shown either advance (24%) or retreat (25%). Beaches between Valparaíso and San Antonio, in central Chile, show seasonal oscillations, but have maintained the general alignments shown on hydrographic charts made in 1875: near the mouth of the Rio Maipo, the sandy coastline has prograded slightly (Del Canto and Paskoff, 1983). Some sectors of the coastline have advanced because of tectonic uplift, or retreated because of subsidence during earthquakes. Charles Darwin observed a sudden land uplift at Talcahuano in the 1835 earthquake, during his voyage on the *Beagle*. In 1960 an earthquake resulted in coastal submergence of up to 2 metres between Concepción and Chiloé Island, with ensuing beach erosion, and associated tsunami scouring of coastal sediments (Weichset, 1963).

In southern Chile the coast is generally steep, with intricate patterns of straits and fiords, and some local progradation related to the longshore supply of sediment derived from glacial drift and glacifluvial deposits. Lobes of glacifluvial sediment have formed in front of melting glaciers, notably behind Chiloé Island, in Skyrig and Otway Sounds, and on the south of Navarino

Island (Araya-Vergara, 1985). Ice cliffs have shown advance alternating with retreat during the past two centuries in the Bay of Lagune de San Rafael (Heusser, 1960), and in Bernardo and Messier Fiords. Deltas are growing at the heads of several fiords, where salt marshes are spreading on to tidal mudflats.

ARGENTINA

In southern Argentina the Tierra del Fuego coastline has receding cliffs cut in a variety of rock formations, including glacial drift deposits, and sediment derived from these has nourished sand and gravel beaches and spits. In addition, rivers draining to this coast have contributed sand and gravel. Similar features are seen along the coast of Bahía Grande, where rivers such as the Gallegos, Chico, and Deseado are delivering sediments to marshy estuarine inlets. The Cabo Blanco coast has receding cliffs, which extend intermittently around the Gulf of San Jorge and on the Valdés Peninsula. Tide ranges of up to 9 metres occur in the Gulf of San Matias, where accreting tidal sandflats are extensive. Progradation continues on the marshy Rio Negro and Rio Colorado deltas and locally on the shores of Bahía Blanca.

Towards Mar del Plata are cliffs in silt and clay, retreating at 2 to 4 metres per year. Erosion is severe at Mar Chiquita, where south of the lagoon entrance the coastline receded up to 160 metres between 1957 and 1979 (Schnack, 1985). Near Buenos Aires, extensive shoals impede navigation in the Rio de la Plata, and onshore drifting from these has prograded sandy shores to the north-west. The upper reaches of the Rio de la Plata are being narrowed by the growth of the Paraná River delta (Iriondo and Scotta, 1979).

URUGUAY

On the Uruguayan shores of the Rio de la Plata, opposite the Paraná delta, the Rio Uruguay is delivering sand to beaches along the coast to Colonia Point. From here to Montevideo there are extensive nearshore sand shoals, from which sand has moved in to prograde beaches (Jackson, 1985). At Montevideo the coast is of hard rocky formations with sandy coves, and port construction has resulted in local beach depletion. To the east are slumping and receding cliffs in Tertiary sediments, with derived beaches of sand and gravel, and spits that have grown eastward, deflecting river mouths. Punta del Este is a recently breached sandy tombolo, the isthmus preserved only by a road and sea wall construction. To the north-east, cliffed promontories and outlets from estuarine lagoons interrupt the sandy beaches, most of which are retreating. Near Castillos, however, dunes driven by south-westerly winds are spilling on to the shore and sand thus supplied has prograded the beach south of the lagoon outlet (Jackson, 1985).

BRAZIL

In Brazil, sandy coastlines are very extensive and during the past few decades they have been generally receding. In the south, the broad sandy barriers seaward of Lagoa dos Patos formerly prograded, but the Atlantic coast of the state of Rio Grande do Sul has recently shown erosion for at least 350 kilometres south of Torres, in an area where man has not interfered with the coastal system (Tanner and Stapor, 1972). There has been minor accretion, with added beach ridges, at the western end of Praia de Fora, close to the Cananéia outlet. Similar features occur in Santa Catarina state and on to Santos, beyond which steep promontories become more frequent between bay beaches. At Caraguatatuba (Fig. 5) the beach prograded after a 1967 downpour in the steep hinterland, where catastrophic erosion fed vast quantities of sand, silt and clay into flooding rivers and thence to the sea (Cruz et al., 1985). The indented coast behind Ilha Grande has only pocket beaches at river mouths, but the Restinga da Marambaia marks the beginning of a series of beaches and barriers that extend past Rio to Cabo Frio, and on these erosion is prevalent (Fig. 11). Muehe (1979) has shown that this modern erosion is a sequel to previous (Holocene) barrier progradation.

North of Cabo Frio is a series of deltas, built by the Paraíba do Sul Doce, São Mateus, Mucuri, Itanhém, Jucururu, Jequitinhonha, Pardo, Contas, Colonia and São Francisco Rivers, each of which has prograded by addition of sandy beach ridges in Holocene times. Some sectors, close to river mouths

Fig. 11 A retreating sandy coastline south of Saquarema, Brazil, where beach erosion threatens to undercut the coast road. Photo: Eric Bird (August 1982)

supplying sand, are still prograding; others are receding (Martin *et al.*, 1980). Although there are extensive steep coasts, active cliffing is limited in Brazil (Cruz *et al.*, 1985). The existence of 'stone reefs' of beach rock offshore between Salvador and Natal implies substantial recession of this coastline, but the time-scale of this retreat has not been determined. In the north-east, mangrove-fringed estuaries interrupt the sandy coastlines, and there have been intricate changes historically near river mouths, with the enlargement of deltaic islands in the mouths of the Pará and Amazon Rivers. As Sioli (1966) pointed out, much of the silt carried down to the sea by the Amazon is swept northwards to be deposited on the swampy coast of Amapa province, and beyond in the Guianas.

FRENCH GUIANA, SURINAM AND GUYANA

The coastlines of these three states are low-lying and swampy, with extensive mangroves behind tidal mudflats, and intermittent sandy beaches, especially near river mouths. The coastline has shown alternating advance and retreat, but the long-term trend has been progradational. Shore accretion and mangrove advance have occurred behind nearshore shoals, but as these migrate westward, the passage of intervening deeper sectors has been marked by erosion of the shore, and the winnowing and accumulation of shelly sand beaches (Wells and Coleman, 1981). A spit is growing westwards at the mouth of the Waini River in Guyana.

VENEZUELA

In Venezuela the Orinoco delta has shown similar progradational features, and swampy shores have encroached on the landlocked Gulf of Paria to the north. Much of the Venezuelan coastline is steep, and there are sectors of active cliffing, notably along the hilly coast of Falcon province, on the Paraguaná peninsula, and the southern side of Guajira peninsula (Ellenberg, 1985). Beach erosion has been extensive near Tucacas, and on the east shore of the Medanos isthmus. Tanner (1975) described a slowly prograding beach ridge plain on the west coast of the Gulf of Venezuela, but noted a general recession of sandy coastlines. Progradation has occurred on the Rio Mitare delta, and on the swampy southern shores of the landlocked embayment of Lake Maracaibo.

CARIBBEAN COLOMBIA

Some cliff recession is in progress along the steep coast of Colombia as on the northern flanks of Cristóbal Colon near Palomino, and beach erosion is prevalent near Barranquilla (Tanner, 1975). Some progradation has continued on deltas, such as the Truando in the Gulf of Uraba. Troll and Schmidt-Kraepelin (1965) documented the growth of a new delta built by the Rio Sinu on the Colombian coast following a change in the position of the river mouth in 1942.

CARIBBEAN CENTRAL AMERICA

The Caribbean coasts of Panama, Costa Rica, Nicaragua, and Honduras include some steep and cliffed sectors, several growing deltas, extensive mangrove swamps, and discontinuous beach and barrier systems. Information concerning changes on these coasts during the past century is scanty, but W. F. Tanner has reported that sectors of continuing accretion were few and localized. Offshore, on the British Honduras reefs, Stoddart (1964) recorded the disappearance of 16 cays between 1830 and 1950, presumably as the result of hurricane erosion: six more vanished in the 1961 hurricane.

CARIBBEAN MEXICO

In Mexico the coast of the Yucatán peninsula consists of rocky limestone and associated carbonate sand beaches which are generally slowly receding, but the low-lying depositional coast which extends almost without interruption from Laguna de Terminos on the southern shore of the Bay of Campeche around to the United States border at the mouth of the Rio Grande now includes only a few local sectors of active progradation related to fluvial deposition. The southern coast, in Tabasco, is essentially a deltaic plain fringed by sandy barriers which enclose lagoon systems between the larger river deltas (Psuty, 1965). Deposition continues, but in recent decades much of the barrier coastline has been receding, and it is only adjacent to river mouths and tidal inlets that sectors of progradation, often only temporary, can be recognized. Thom (1969) illustrated beach ridge truncation on the eastern part of the north shore of the Isla del Carmen, and progradation of beach ridges to the west, alongside Borra Principal, an inlet to the Laguna de Terminos. W. F. Tanner has reported continuing sandy progradation locally, adjacent to artificial structures, for example west of the mouth of the Grijalva River in Tabasco and east of the river mouth at Alvarado. In general, the beach-ridge plains and barriers that previously prograded on the Caribbean coast of Mexico (including Caba Roja and the barrier islands seaward of the Mexican Laguna Madre) are now being cut back by marine erosion, and in Veracruz the retreating coast is backed by extensive parabolic dune systems, moving inland. Tanner and Stapor (1971) deduced that there was a late Holocene abundance of sand supply, when the beach-ridge plains of Tabasco were built, followed by a phase when the sand supply diminished, leading to the onset of erosion. The seaward margins of dunes are now cliffed, and trees that once grew on former beach ridges now stand out in the surf.

USA—GULF COAST

The Gulf Coast of the United States is almost entirely low-lying, with sandy barriers in Texas, marshy and deltaic shores in Louisiana, more sandy barriers eastward as far as Apalachee Bay in Florida, then marshy shores, with mangroves, and minor sandy areas in southern Florida. Shepard and Wanless (1971) indicated only limited sectors of progradation here during the past century, and several of these were the outcome, directly or indirectly, of human interference; otherwise erosion has been prevalent, with episodes of rapid recession during occasional hurricanes.

Details of coastline changes in Texas have been summarized by Morton (1977), who noted that net accretion over the past 125 years had been confined to parts of Matagorda Island and Padre Island within zones of longshore drift convergence. In the past decade, 70% of the Texas coastline has retreated, the land losses averaging about 160 hectares annually. Fluvial sediment from the Rio Grande, at the Mexican border, is dispersed by wave action, and apart from local accretion adjacent to breakwaters at Brazos Santiago Pass the sandy coastline has not advanced here during the past century. The barrier islands of the Texas coast, which prograded with the addition of successive beach ridges earlier in Holocene times, now have generally stable or retreating coastlines. On Matagorda Peninsula, Wilkinson and McGowen (1977) found erosion rates from 1856 to 1957 to be 30–40% greater than average coastline recession over the past 900 years, determined by dating shore oysters deposited by overwash and now buried beneath transgressive barrier sands. They attributed this acceleration to the impacts of human activities. During hurricanes, rapid recession occurs. In 1961 the coastline of the Matagorda peninsula was cut back up to 250 metres during Hurricane Carla, but subsequently much of this loss was restored. Accretion has continued alongside protruding jetties built at Aransas Pass, and at Galveston Pass, where the coastline has advanced a kilometre since 1894. Most of the Texas rivers discharge into lagoons behind the sandy barriers: the Colorado built a delta across the lagoon behind Matagorda peninsula in the 1930s, and a canal cut in 1936 to provide an outlet to the Gulf of Mexico has since become an area of further deltaic growth (Wadsworth, 1966). The Brazos River, by far the most prolific sediment-yielding river in Texas, is exceptional in that it has built its delta out naturally into the gulf. Between 1858 and 1938 this arcuate delta prograded up to 1.2 kilometres seaward, but an artificial outlet cut to the west has since become the site of a new delta, which grew about 1.6 kilometres seaward in 20 years, while the earlier delta was cut back and dissected by wave action.

Morgan's (1963) map of coastline changes in Louisiana between 1812 and 1954 (Fig. 12) showed sectors of advance alongside the mouth of Sabine Pass, at the mouth of Calcasieu River, on a lobate shore south-west of Vermilion Bay, and on the modern subdelta of the Mississippi, where the digitate silt jetties have lengthened: South-west Pass, for instance, has grown

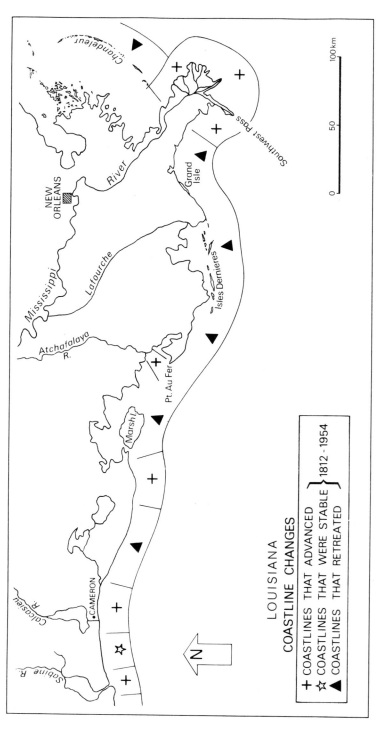

Fig. 12 The pattern of coastline advance and retreat in Louisiana, obtained by comparing maps made in 1812 and 1934 (after Morgan, 1963)

about 8 kilometres during the past century. Intervening sectors, including parts of former Mississippi subdeltas, showed active retreat between 1812 and 1954. In recent decades, high rates of erosion have been prevalent on parts of the deltaic coastline, flood control works having diminished fluvial sediment yield, and reduced its dispersal over the coastal zone, which is here subsiding at over a metre a century. In 1971 the National Shoreline Study mapped gradual erosion along much of the Louisiana coast, and severe erosion on Grand Isle, a barrier island off Barataria Bay, and on parts of the shoreline of Lake Pontchartrain, especially the south-east, at Lincoln Beach. There had been local accretion against artificial breakwaters, as at Belle Passe and the outlet of the Empire to Gulf of Mexico waterway. Otherwise the main sectors of stable or accreting coastline were east of Sabine Pass, along Ocean View Beach, from Calcasieu Lake outlet to Rutherford's Beach, west of Chenière au Tigre, and around the modern Mississippi delta (Shepard and Wanless, 1971). There has been rapid progradation in recent years around the mouth of the canalized Atchafalaya River, formerly a Mississippi distributary (Heerden and Roberts, 1980).

East of the Mississippi it is possible to detect minor changes on spits and barrier coastlines during the past century. A new barrier island has developed off Ono Island at the mouth of Perdido Bay, and the beach extending eastwards to Pensacola Bay (Gulf Beach) has shown sustained accretion, averaging 60 metres, during the past century as the result of the input of sand dredged from the entrances to these two bays (Price, 1975). East of Pensacola Inlet, Santa Rosa Island has grown by accretion since it was surveyed in 1856–9, but over the same period Pensacola Beach has lost about a metre a year. Coastline change farther east is difficult to measure from comparisons of surveyed data. Despite the fact that the seaward margin between Phillips Inlet and Miramar Beach shows a cliff 6 to 15 metres high, cut in pre-Holocene sands, field evidence that would otherwise be interpreted as indicating active coastline retreat, Stapor (1975) was unable to demonstrate any recession here in recent decades. Shell Island Spit near Panama City has shown historical growth, and comparisons of charts made in the 1860s with the modern outlines of St Josephs Point, Cape San Blas, St Vincent and St George Islands all show patterns of gain and loss, the limited zones of continued accretion being where waves were weakened by refraction across offshore shoals (Stapor, 1971). Cape San Blas showed the most rapid beach retreat, 11.2 metres per year between 1875 and 1942 at the lighthouse, while the western shore of St Joseph's Point and the seaward front of Dog Island both receded about a metre a year. St Vincent Island is a good example of a barrier island with a formerly prograded sandy coastline marked by numerous parallel beach ridges, on which there has been a transition to erosion and coastline retreat within the last two or three centuries (Tanner and Stapor, 1972).

There have also been changes in the configuration of barrier islands north and south of Tampa Bay, notably on Lacosta Island, where a cuspate foreland was built up in the lee of an offshore shoal, which emerged as a lunate islet during a hurricane, and at Blind Pass, between Sanibel and Captiva Islands, where longshore spit growth between 1883 and 1965 deflected an inlet, and finally sealed it off. Beach recession of up to 0.6 metres per year typified a 24 kilometre sector of the barrier near Sarasota Bay between 1957 and 1973 (Banks, 1975).

Delta growth has continued in Mobile Bay, where the Tenshaw and Mobile Rivers advanced their combined delta coastline by an average of 3.2 kilometres between 1890 and 1960, and in the lagoon behind St George Island, where the Apalachicola is enlarging its delta. Marshy sectors of the Florida coast on either side of the mouth of the Suwannee River have shown alternations of advance, with vegetation spreading forward on accreting sediment, and retreat, especially during hurricanes. The same applies to the mangrove-fringed south-western shores of peninsular Florida, which have changed intricately in recent decades, with some sectors of shelly beach accumulation, as on Cape Sable.

CARIBBEAN ISLAND COASTS

Caribbean island coasts have been generally stable or slowly eroding over recent decades, progradation having been confined to river mouth areas and swamp-fringed embayments (Tanner, 1975). In Cuba beach erosion is widespread, particularly on the north coast near Havana. In Jamaica, Wood (1976) reported evidence of accretion on parts of the south coast in the interval between Liddell's 1888 survey and maps and air photographs produced in the 1960s. Progradation has occurred at river mouths, and where westward drifting of fluvially supplied sediment has nourished beach or swamp advance; locally, materials derived from cliff or reef erosion have enlarged beaches. Hunts Bay in Kingston Harbour has an advancing swampy coastline, and despite recurrent hurricane damage the Palisadoes barrier spit at the mouth of Kingston Harbour has maintained its position by the addition of material drifting from the mouth of Hope River, to the east. On the steeper north coast of Jamaica the river basins are smaller, and deposition has been localized within river-mouth inlets and embayments.

In Hispaniola, Alexander (1985) described receding cliffs, and commented on the progradation of mangrove sectors receiving an increased fluvial sediment supply derived from areas of severe soil erosion in the hinterland. Erosion has become extensive on beaches in Puerto Rico, partly as a consequence of engineering works, such as breakwater and causeway construction, partly as a result of beach sand mining (Morelock, 1978). Cliffs are retreating in limestone and soft sedimentary outcrops, and erosion is locally rapid on the south coast of the island, east of Ponce (Morelock and Trumbull, 1985). The small islands of the Lesser Antilles have receding cliffs in limestone and volcanic rock, and beaches in many places show evidence of actual or incipient erosion (Deane, 1985). On the east coast of St Vincent, beach erosion has accelerated following the extraction of sand and gravel from the foreshore for use in building and road-making, while on the west coast of Barbados beach rock has become widely exposed as the result of erosion of the sandy beaches (Bird, Richards and Wong, 1979). Cliffs in Miocene sands and clays on the west coast of Trinidad have retreated up to 2 metres per year in recent decades. By contrast the Bahamas, sited on wide shallow banks of calcareous sediment, have prograding beaches where biogenic sand is being carried shoreward (Craig, 1985).

USA—ATLANTIC COAST

The extensive sandy Atlantic coastlines of the United States have shown only limited sectors of sustained progradation during the past century, the general tendency being recession, often accompanying landward migration of transgressive barrier formations (Hayes, 1985; Kraft, 1985; Fisher, 1985). The Miami coastline retreated about 150 metres between 1884 and 1944, about half the recession occurring during a single hurricane in 1926. Predominant southward drifting of beach material has resulted in local accretion alongside breakwaters, as at Lake Worth Inlet, near Miami, since the jetties were built in 1918–25. There was up to 450 metres of beach progradation adjacent to the northern jetty by 1959, and corresponding erosion to the south. W. F. Tanner reported that a short segment of sandy coast north of Daytona Beach has advanced at about a metre a year since the beginning of this century.

On the barrier islands of the Georgia coast the pattern of historical change since 1897 has been irregular, with local alternations of advance and retreat (Oertel and Chamberlain, 1975). Sustained or net accretion has generally been localized on spits and cuspate forelands, in contrast with the broad seaward advance of the ocean coastline that occurred earlier in Holocene times, commemorated by successively formed parallel beach ridges. Fluvial sediment, notably sand from the Savanna River, has augmented the material locally available for coastal deposition, but direct nourishment of beaches from fluvial sources has been very limited on the Atlantic coast of the United States.

The major cuspate forelands of the Atlantic coast have each migrated southward during the past century. Cape Kennedy in Florida had erosion of its north-eastern flank and accretion along the southern shore, continuing the long-term trend indicated by its beach ridge pattern. Cape Romain on the Carolina coast lost up to 1200 metres on its eastern flank and gained a new sandy barrier on its southern shore between 1886 and 1963. Cape Fear lost up to 500 metres on its eastern shore and gained up to 200 metres on its southern shore between 1849 and 1929, and Cape Hatteras showed the same pattern of receding eastern coastline and prograding southern coastline when its outline on an 1852 chart was compared with more recent maps (1872, 1917, 1939) and air photographs (1945, 1953, 1962) (Fisher, 1980): a more recent (1980) map by the National Shoreline Study shows a readvance of the Cape southwards. By contrast, Cape Lookout has been widened, its land area increasing from about 200 to about 400 hectares between 1866 and 1955 (Pierce, 1969), but erosion on its southern side prompted Dolan *et al.* (1980) to predict that the lighthouse will be undermined by 1991.

Intervening sectors of barrier coastline have generally receded, although there have been changes on either side of lagoon entrances unconfined by artificial structures, where transverse currents act as a natural breakwater, impeding longshore sediment flow. The Carolina coast has had a long history of enlargement, reduction, closure, and reopening of such entrances (Dunbar, 1956). Measurements on 165 equally spaced transverse shore sections between Hatteras Inlet and Cape Lookout indicated net accretion during the past century, with notable land gains at Cape Lookout and alongside Ocracoke Inlet: Pierce (1969) ascribed this sand accumulation primarily to shoreward drifting from the sea floor. J. McHone reported fluctuations of the sandy coastline at False Cape, in southern Virginia, which may be related to variations in sand input from an adjacent migrating submarine sand ridge. In a survey of historical erosion and accretion along the central east coast of the United States (Goldsmith *et al.*, 1978) found evidence of substantial accretion to the north of Cape Hatteras and near the Virginia-Maryland border, but erosion predominated elsewhere. Cape Henlopen, on the southern side of the mouth of Delaware Bay, has shown pronounced spit growth northward accompanying recession of the Atlantic coastline since 1842 (Fig. 13). Within Chesapeake Bay, Hunter (1914) found evidence of land gains around river mouths and losses in intervening cliffed sections when he compared surveys made in 1847–8 and 1910. Deposition at river mouths, probably accelerated by soil erosion in the catchments, has continued, promoting the seaward advance of salt marshes and succeeding woodland in tributary creeks (Froomer, 1980).

Coastline recession has also predominated in New Jersey, and continues, despite extensive groyne and wall construction. Dolan *et al.* (1978) analysed erosion and accretion on the sandy barrier islands of southern New Jersey from 1930 to 1971, and found local progradation of up to 30 metres/year and local recession of up to 20 metres/year, the largest changes being on either side of tidal inlets, notably Hereford, Carson and Brigantine Inlets. Caldwell (1966) noted that part of the eroded sediment has moved north towards Sandy Hook, but much of it has been carried away offshore. Changes have occurred on the spit at Sandy Hook since 1840, but the erosional loss has balanced accretional gain here (Shepard and Wanless, 1971): in recent years the coastline has been armoured, and some sectors of eroded beach replenished artificially (Nordstrom and Allen, 1980). McCormick (1973) used transverse shore profiles to deduce the recent history of advance and recession of beaches on south-eastern Long Island, with supplementary evidence from comparisons of historical surveys (1838, 1891, 1933, 1956). Progradation has occurred in the embayment east of East Hampton and erosion to the west: the opening of tidal inlets has resulted in the interruption of longshore

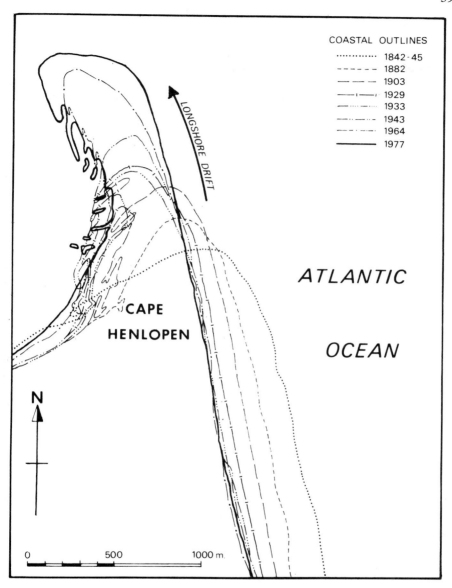

COASTAL OUTLINES

············	1842-45
------	1882
— — —	1903
—ᵢ—ᵢ	1929
—·····—	1933
—··—··—	1943
—·— ·—·	1964
———	1977

LONGSHORE DRIFT

CAPE

HENLOPEN

ATLANTIC

OCEAN

N

0   500   1000 m.

Fig. 13 Intermittent northward growth of the spit at Cape Henlopen, on the southern side of the mouth of Delaware Bay, has accompanied recession of the Atlantic coastline to the south. Sediment eroded from the retreating ocean coastline has drifted north to be incorporated in the growing spit. Similar outlines from successive maps and charts can be drawn for several such spits on the Atlantic coast of the United States, but in some cases (e.g. Cape Hatteras) the growth has been southward. Based on data supplied by Craig Everts

drifting by strong transverse currents, with updrift accretion and downdrift erosion, much as if an intercepting breakwater had been built. Rates of cliff recession have influenced rates of adjacent beach erosion here, the coastline maintaining an almost straight outline in plan. Spit growth has deflected Fire Island Inlet, Long Island, westwards between 1834 and 1946, without much change in the coastal alignment. On Rhode Island Fisher and Simpson (1979) found that coastline recession had average 0.7 metres per year from 1938 to 1975.

On the south coast of Martha's Vineyard Island, Massachusetts, Ogden (1972) traced a long history of coastline recession, notably on the barrier at Katama Bay, which was cut back more than 880 metres between 1776 and 1969, with onset of cliff erosion in the glacial drift terrain to the rear of Wasque. Kaye (1973) produced a detailed map (1 ' : 24,000) of coastline changes along the coast of Martha's Vineyard Island from Des Barres' 1776 charts in comparison with more recent historical maps and air photographs. Recession of cliffed coasts in glacial drift predominated, but there had been local beach accretion. There was growth of beach ridges on Monomoy Island, to the north, between 1899 and 1955, and at Chatham there has been intermittent growth on the recurved spits bordering the migratory Nauset inlet between 1872 and 1965. On Cape Cod, cliff recession of up to 54 metres between 1887 and 1957 was partly compensated by accretion on the spit at Princetown (Zeigler et al., 1964). There are receding cliffs of glacial drift on other parts of the New England coast, including the drumlins in Boston Harbour, where unpublished surveys by J. J. Fisher and P. Riegler showed an excess of erosion over deposition between 1938 and 1977, accretion being generally at the southern ends of the islands, in response to strong northerly wind and wave action. Farther north, in Maine, the harder rocky shores have shown very little change in the past century (Shepard and Wanless, 1971).

CANADA—ATLANTIC COAST

Cliff recession has been locally rapid in eastern Canada, in places where marine erosion is working upon glacial drift deposits, as on parts of the coastline on the Bay of Fundy and the Gulf of St Lawrence. On Prince Edward Island there has been rapid recession of cliffs cut in Permo-Carboniferous shales and sandstone, especially on the north coast near Cape Tryon, but the harder rock outcrops which dominate much of the coast of Nova Scotia, Quebec, Labrador and Newfoundland have changed very little during the past century. Where glacial drift cliffs are eroding, material is available for nearby deposition on beaches and spits, and for accretion in marshes, as exemplified towards the head of the Bay of Fundy.

On the western and southern shores of the lower Gulf of St Lawrence there are extensive sandy beaches and barriers. Progradation has been rare, but the extremities of barrier islands often show lateral growth of spits, as at the eastern end of Hog Island, on the north coast of Prince Edward Island.

Armon and McCann (1977) found that the Malpleque barrier coastline had retreated by up to 100 metres between 1845 and 1955, the recession being accompanied by lowering of the nearshore sea floor. On the north shore of Prince Edward Island, J. W. Armon reported general coastline recession and suggested that continuing submergence in the southern Gulf of St Lawrence had contributed to this: estimates vary, but at Charlottetown tidal records indicated submergence at the rate of 2.6 millimetres per year since 1930. Lateral growth has taken place on spits and barrier islands on New Brunswick coast, notably in Miramichi Bay, but these sandy features are being driven landward to expose marsh and lagoon deposits on their seaward margins; in some places there are peat cliffs (Owens, 1974). Bryant (1979) concluded that historical retreat and spit growth in Kouchibouguac Bay were a response to wave climate effects rather than to catastrophic (storm surge) events. In the Maritime Provinces generally, cliffs have receded up to 3 metres per year in exposed sections cut in unconsolidated Quaternary deposits (Owens and Bowen, 1977).

The Magdalen Islands, out in the gulf, are extensive sandy depositional features, linking rocky islands and enclosing shallow lagoons. Their sandy coastlines have varied historically, and in recent decades erosion has been prevalent. The beaches are generally retreating, the exception being at the northern end, where comparisons of maps made in 1833, 1890, 1898, and 1934, and air photographs taken in 1952 and 1977, indicated substantial accretion, attaining a maximum of 13.5 metres per year at Pointe de L'Est.

Locally on the emerging north shore of the St Lawrence estuary the Quebec rivers are building small deltas, and providing sediment for the continued growth of beaches and spits (Dubois, 1980) and accretion on boulder-strewn salt marshes (Dionne, 1979). Much of this coastline is rocky, and has changed little in recent decades. Shore ice develops and lasts up to four months in winter, changes occurring on beaches and marshes when it breaks up in the spring thaw. In James Bay, south of Hudson Bay, continuing emergence has led to rapid progradation of a low-lying coastline (Martini, 1981). J. M. M. Dubois reported that the islands of Saint-Pierre and Miquelon are rocky, with intervening sandy tombolos, which are being eroded, especially during storms. In 1759 a storm breached the tombolo between Miquelon and Langlade, forming a strait which was 1800 metres wide in 1763. Subsequently the tombolo has been repeatedly rebuilt, and recurrently storm-breached.

CANADA—ARCTIC COAST

Sea ice and fast ice restrict the operation of marine processes to the brief summer thaw in Arctic Canada, but there are receding cliffs cut in glacial drift on Victoria Island, King William Island and Cape Bathurst, and limestone cliffs on Baffin Island. Active periglaciation is yielding angular gravel for beach and spit construction on the shores of Devon and Somerset Islands

(McCann, 1973), and there are areas of accumulation of tundra marsh and mudflat east of Foxe Basin (King, 1969). Little change has occurred along the fiord coasts of Baffin Bay, but Ellesmere Island has sectors of ice shelf coast which have shown both advance and recession during the past century. In some areas isostatic uplift following deglaciation has led to coastline advance by the emergence of parts of the former sea floor (Andrews, 1973). In the north-west, the Mackenzie River is delivering sediment to prograde its deltaic coastline and nourish beaches and barriers on the shores of the Beaufort Sea (Mackay, 1963a).

GREAT LAKES

The coastlines of the Great Lakes are included in this review because these lakes have shown climate-related changes of level, with substantial rises and falls independent of the much smaller world-wide fluctuations of sea level during the past century. They have minor tides (up to 10 centimetres) and occasional seiches and storm surges attaining 2 metres amplitude, accompanying seasonal variations, early summer levels being up to 1.2 metres above those of early winter. There have also been longer-term fluctuations: Lakes Michigan and Huron stood up to 0.6 metres above datum (176.5 metres above sea level) in the 1880s, but since 1890 have been generally below datum, with brief high stands in 1906, 1917, 1926, 1952–3, and 1973 (Larsen, 1973). In addition, tilting due to isostatic rebound is raising the coastlines to the north-east, especially in Lakes Huron and Ontario, relative to those of the south-west, notably in Lake Erie, by a differential of 0.3 metres per century.

R. D. Gillie reported that progradation of Great Lakes coastlines has been restricted to growing portions of spits (such as at Pointe aux Pins and Long Point in Lake Erie), and to longshore drift accumulation at the ends of bays (as at Wasaga Beach in Georgia Bay on Lake Huron, and Burlington Beach at the western end of Lake Ontario). Coakley (1976) gave an account of historical changes in the growth of a sandy foreland at Point Pelée, western Lake Erie. Erosion has been very extensive. The US National Shoreline Study in 1971 identified 66 beaches on the coastlines of the Great Lakes which then had critical erosion problems: they represented 343 kilometres, 5.8% of the total coastline (5887 kilometres). A further 1672 kilometres (28.4%) were also eroding. Already 608 kilometres (10.3%) were protected by artificial structures. For example erosion of the north shore of the Presque Isle peninsula, a recurved spit on the south side of Lake Erie, since 1872 has resulted in the building of walls and groynes, and more recently artificial beach replenishment. Carter (1978) found coastline recession of up to 2.8 metres per year between 1876 and 1973 on the Lake County section of southern Lake Erie, with variations related to the introduction of engineering structures, including breakwaters that intercepted longshore drifting to give accretion to the west and erosion to the east.

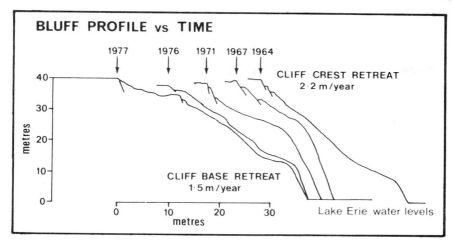

Fig. 14 Repeated surveys showing variations in the profile of a retreating cliff at North Shore Bluffs, Point Bruce, Lake Erie, between 1964 and 1977 (Quigley and Di Nardo, 1980)

Larsen (1973) correlated phases of beach retreat on the shores of Lake Michigan between 1872 and 1955 with episodes of high lake level, supporting the Bruun theory (see page 169) of shore profile response to submergence (Schwartz, 1967). Olson (1958) related phases of sandy coastline advance and beach ridge and foredune construction to episodes of falling lake level, but over the past century the sandy coastlines of Lake Michigan have shown net retreat.

Although independent of oceanic fluctuations, the Great Lakes coastlines conform with the world-wide pattern of predominating beach erosion. Cliffs cut mainly in glacial drift have also shown recession, accelerated during phases of high lake level. W. R. Buckles mapped cliff recession at 118 locations on the eastern and western shores of Lake Michigan, averaging 1 to 2 metres per year. At Scarborough Bluffs, on the north shore of Lake Ontario, Bryan and Price (1980) have documented cliff retreat, due partly to recurrent slope failure, of up to 0.5 metres per year. The profiles of a receding bluff at Point Bruce, on the north shore of Lake Erie, from 1964 to 1977 are shown in Fig. 14.

GREENLAND

Ice coasts are extensive in northern Greenland, especially along the shores of Melville Bay, but the extent to which these have advanced or receded during the past century is not known. Calving and melting at the ice margin is at least partly compensated by continuing seaward glacial flow, and the alignment of an ice coast is difficult to monitor in the absence of fixed topographic datum points. Advance and retreat of an ice coast are not simply related to climatic cooling and warming: a mild winter could result in the seaward advance of a glacier or ice sheet that had received increased snowfall, whereas a cold dry winter may reduce the rate of ice movement.

In southern Greenland there has been recession of the snouts of valley glaciers within fiords on rocky coasts within the past century (Ahlmann, 1949), and on the fiord coast, bordering Davis Strait periglacial solifluction has generated gravelly deposits which are being reworked to form beaches and barriers, for instance at the eastern end of the island of Disko (Nielsen, 1969). The southern and western coasts of this island have high, receding cliffs. Ninety per cent of the Greenland coastline consists of hard rock outcrops, on which changes are very slow, except for local seasonal plucking by the ice-foot, the ledge of ice that forms each winter along the sea shore. Glacifluvial fans and deltas are prograding at the heads and along the sides of several fiords. The sequence of tectonic movements in Greenland has been complex, but some emergence through uplift has been reported over the past 40 years (Nielsen, 1985).

ICELAND

Iceland has some recently formed volcanic coastline, with lava and ash deposits into which cliffs are being cut back by marine erosion. The island of Surtsey, built by volcanic activity between 1963 and 1967, developed cliffs up to 32 metres high which had retreated by up to 140 metres in 1967–8, and have since receded by up to 40 metres per year (Norrman, 1980). Eroded lava fragments have accumulated locally in boulder beaches, and a large depositional foreland of volcanic sand and gravel has grown at the northern end of the island (Fig. 15).

On the south coast of Iceland there are extensive beaches and barriers of black basaltic sand built up by wave action, which has been reworking the sandur (glacial outwash) plains. Sand has been deposited on either side of the rocky peninsula of Ingolfshöfdi, which was shown as an outlying island on a map dated 1794 (King, 1956). The sandur, extending out beneath the sea, built up here as a result of massive discharge of glacifluvial sediment from ice suddenly melted by volcanic eruptions of Öraefajökull, notably in 1362 and 1727. A former cliffed coast was thereafter fronted by a broad area of very shallow water, whence wave action has derived sandy material and carried it shoreward. Progradation is still in progress along this part of the coast, and there are some other sites of coastline progradation related to sandur accretion on the north-west coast of Iceland, for example in the Axarfjördur (Bodéré, 1985).

NORWAY

Much of the coastline of Norway is steep and rocky, indented by fiords, and changes during the past century have been negligible where hard rock

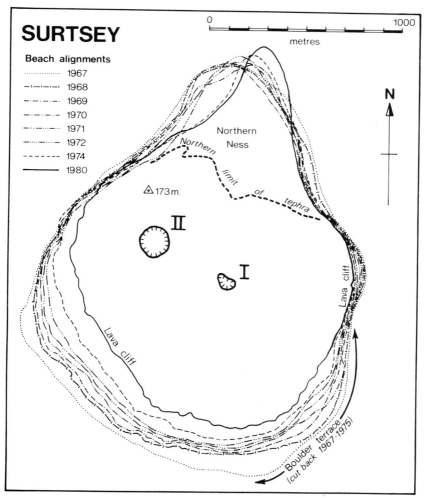

SURTSEY

Beach alignments
```
................ 1967
—·—·—·—· 1968
—·—··—·—· 1969
—···—···— 1970
—····—····— 1971
—·····—·····— 1972
— — — — 1974
———— 1980
```

Northern Ness

Northern limit of tephra

△ 173 m.

II

I

Lava cliff

Lava cliff

Boulder terrace 1967-1975)
(cut back

Fig. 15 Maps made from successive air photographs show the changing outline of Surtsey, a volcanic island built by eruptions in 1966–7 off southern Iceland. Cliff recession on the more exposed southern coastline has been accompanied by longshore drifting northwards on the east and west coasts, and the growth of a depositional foreland, the Northern Ness, which has been migrating eastward (Norrman, 1980; Calles et al., 1982)

formations outcrop along the shore. In many places the striations produced by the passage of glacial ice are still clearly visible on rocky shores, indicating a lack of marine erosion over the past few thousand years. Active cliffing is very limited in Norway. Isostatic uplift following deglaciation has raised the coasts, especially in the south-east, where the land is still rising at about 30 centimetres per century in the Oslo region. Accretion of mudflats and seaward advance of salt marshes as the head of inlets and embayments around the Oslo Fiord has been aided by this continuing emergence (Klemsdal, 1985). In southern Norway sandy beaches backed by dunes have been derived from glacifluvial sands associated with morainic deposits, notably the Ra moraine which marks the ice limit in Scandinavia about 10,500 years ago. The sands have been reworked by fluvial action where the sources lie inland, and by marine processes where the glacial deposits are out on the sea floor, and thence delivered to the coastline. Thus the beach at the head of the Lyngdal Fiord is still receiving sand brought down by the Lyngdalselva River and continues to prograde, while beaches at Brusand to the west, and on the coast of Karmoy Island, are being maintained by sand washed in from the sea floor. By contrast, the sandy beaches of the Jaeren region, south of Stavanger, are backed by cliffs cut into backshore dunes, as is the curving sandy beach at Mandal in southernmost Norway. Such erosion takes place mainly during storms, and its prevalence may be related to a decline in natural beach replenishment, especially from offshore sources. Deposition from rivers draining into fiords has produced some local progradation, but much of this fluvial sediment is carried offshore. Per Bruun (1962) reported that occasional subaqueous mass movements within fiords had resulted in episodes of erosion on bordering shores.

The coastline of Spitsbergen is mainly rocky, but locally beach sediments have been supplied from melting glaciers and icebergs. Moign and Guilcher (1967) described the evolution and historical growth of a cuspate spit at Sars, due to convergent drifting sand and gravel on the shores of a strait on the west coast, in the lee of Prince Karl Island.

SWEDEN

The coastline of southern Sweden has changed little during the past century, except for localized deposition on beaches and spits, the progradation of marshes in sheltered inlets and embayments, and some cliff recession. Rudberg (1967) estimated that the cliffs of limestone and marl on the west coast of the island of Gotland were receding at 0.4 to 0.6 centimetres per year. Only in Skåne, the southernmost province, is coastal erosion extensive, and here several beaches are being cut back, for example at Löderup, where the coastline retreated about 100 metres between 1818 and 1965, at Ystad, at Falstad, and at Falsterbo (Lindh, 1976).

L. E. Åse and J. O. Norrman reported that a comparison of precise levelling in 1886–1905 and 1951–66 indicated a pattern of land uplift increasing northwards through Sweden, with maximum rates along the coast of the Gulf of Bothnia between Umeå and Luleå (Fig. 16). This had resulted in an advance of the coastline, especially where coastal waters were shallow. It represents the continuation of a long-term trend. As the land rose, harbours

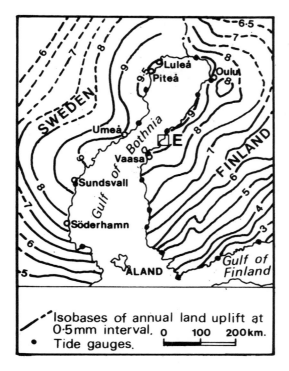

Fig. 16 The pattern of land uplift in northern
Scandinavia, due to continuing postglacial isostatic recovery, has resulted in active emergence
around the shores of the Gulf of Bothnia (after
Jones, 1982). Area E is shown in Fig. 18

have been abandoned, and in places runic stones which around AD 1000
marked navigable channels are now found in fields far inland; watermarks
cut on cliffs to record mean sea level in the eighteenth century have been
raised to higher levels. On the Bothnian coast, land uplift at 1 centimetre
per year has resulted in an advance of the coastline averaging a few metres
per century, but on gently shelving shores it has attained 100 metres per
century. Coastal advance has also been aided by deposition of sediment on
the emerging shores, especially near river mouths, and by the accumulation of
peat under swamp vegetation colonizing shallow, sheltered waters. However,
most rivers have been dammed to provide reservoirs for hydroelectric power
generation, and their sediment yields have greatly diminished. Despite an
emergence rate of almost a metre per century, the sandy coastline of the
Indalsälven delta on the Bothnian coast is now retreating. The advance of a
coastline because of isostatic uplift has also been documented in eastern
Svealand, and on Åland, by Åse (1970).

FINLAND

Continuing emergence is also a factor in the modern evolution of the Finnish coastline (Jones, 1977), especially in the north-west, where uplift is proceeding at a rate of about 1 centimetre per year: it diminishes south-eastwards to less than 2 millimetres per year in the eastern Gulf of Finland (Alestalo, 1985). Sandy beaches are usually related to the erosion and shore-ward drifting of sea floor esker deposits. Sand deposition has formed success-ively emergent beach ridges near Oulu (Helle, 1965), and the seaward spread of fresh and brackish marshes is in evidence around Vaasa (Fig. 17). In this locality, numerous islands have emerged, enlarged, and coalesced as the result of uplift, and some harbours have been abandoned because of the shallowing of their formerly navigable approaches. The rate of uplift is sufficient for the coastline to have advanced up to a few hundred metres locally in sectors that were shallowly submerged a century ago (Fig. 18); in some cases, the new land is simply the emerged sea floor, in others it has acquired a veneer of accumulated sediment in the course of emergence.

In southern Finland, where the rate of emergence is much less, there has been local progradation as the result of marsh encroachment, and growth of sand and shingle spits. However, the rivers are small, and their sediment yield meagre, and it is only where sediment is derived from sea floor and coastal outcrops of unconsolidated glacial drift that beaches are still prograding. On rocky shores in the archipelago of south-west Finland the

Fig. 17 Continuing emergence of the coastline of north-west Finland near Kala-joki, the outcome of postglacial isostatic uplift of the land in northern Scandinavia, is marked by a seaward advance of marshland. Photo: Eric Bird (May 1970)

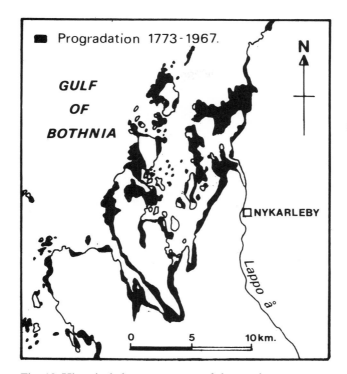

Fig. 18 Historical changes on part of the north-west coast
of Finland near Nykarleby between 1773 and 1967, as a
result of continuing isostatic uplift of the land (cf. Fig. 16)
(after Jones, 1982)

persistence of clearly defined glacial striations testifies to a very slow rate of
weathering and marine erosion during the past few thousand years. Erosion
has been more rapid where a cliffed coastline intersects eskers which have
emerged above present sea level (Granö, 1981).

BALTIC USSR, ESTONIA, LATVIA AND LITHUANIA

On the southern coast of the Gulf of Finland erosion is predominant, except
for minor accretion on beaches near Narva. Glacial drift cliffs are receding
near Tallinn and around islands north of the Gulf of Riga, but beach accretion
has occurred during the past century on the southern shores of the Gulf of
Riga, an area of drift convergence, and on the Kurzeme peninsula, where
sand drifting northward from the Sambian peninsula is accumulating on and
around Cape Kolkas (Gudelis, 1985). V. P. Zenkovich reported that some
formerly prograded beaches in the Gulf of Riga are now retreating. Erosion
has also predominated along the beach and barrier coastlines of Latvia and

Lithuania, south of Klaipeda and south-west of Kaliningrad, but on the Sambian peninsula there are receding cliffs cut into glacial drift deposits (Gudelis, 1985).

POLAND

The Polish coastline has also been generally receding (Borówca, 1985). The recession is obvious in cliffed sectors, mainly of glacial drift, which make up about a quarter of the coastline: on the Orlowo cliff, west of Danzig, map studies show a retreat of about 150 metres between 1847 and 1959. Zenkovich (1967) quoted measurements of cliff recession averaging a metre a year, with up to 5 metres in a stormy year, and mentioned a monastery east of Kołobrzeg, built 1.5 kilometres inland in the thirteenth century, and now undermined by the sea, indicating that cliff retreat over six centuries has averaged 2.5 metres per year.

Sandy beach and barrier coastlines are extensive, and longshore drifting diverges eastward and westward from the Kołobrzeg region. Eastward drift of sand continues to extend the Hel spit out into the Gulf of Danzig, and westward drift into the Gulf of Pomerania has widened the sandy barrier enclosing Szczeciń Lake. Beach progradation has continued near the mouth of the Vistula River, but elsewhere beaches are in retreat, and cliffed backshore dunes are typical. Local accretion has occurred alongside man-made structures, as at Władysławowo, near Cape Rozewie, where harbour breakwaters built in 1933–7 have intercepted sand drifting from the west to advance the coastline more than 200 metres while beach erosion developed to the east. Progradation on the shore of the Gulf of Pomerania has been shaped partly by the jetties built alongside the south of the Swina Inlet, which stand central to a sandy foreland that has advanced at about 4 metres per year since 1800; but this has been an area of sandy beach accretion since the late seventeenth century.

EAST GERMANY AND BALTIC WEST GERMANY

The East German coast shows similar features to those of Poland. Zenkovich (1967) has traced the eastward migration of Darss foreland since the end of the seventeenth century, the retreat of the western flank being accompanied by an advance of the eastern coastline. Cliffs cut in glacial drift are retreating at about a metre per year, and chalk cliffs on the island of Rügen rather more slowly. In the West German sector of the Baltic coast there is similar cliff recession, and eroding beaches alternate with prograding sectors at Timmendorf, Pelzerhaken, and on the island of Fehmarn, and with growing spits at Graswarder, near Heiligenhafen, and Latseninsel at the mouth of the Schlei River (Gierloff-Emden, 1985a). Voss (1970) has shown that the cuspate foreland at Geltinger Birck, which earlier prograded, has in recent centuries been subject to submergence and dissection.

DENMARK

Historical evidence of coastline changes in Denmark during the past two centuries has been reviewed by Bird (1974), using the series of maps published by the Royal Danish Academy of Sciences between 1766 and 1825 (reprinted by the Danish Geodetic Institute in 1956), which cover the coast of Denmark at a scale of 1 : 120,000 and, more particularly, by the original working maps (dated surveys at 1 : 20,000 made between 1762 and 1805) kept in the archives of the Geodetic Institute in Copenhagen. These were compared with successive editions of maps published at a scale of 1 : 20,000 (and more recently 1 : 25,000) by the Danish Geodetic Institute, and with air photographs taken within the past 20 years. On this basis it was possible to identify sectors of the Danish coast (total length just over 7400 kilometres) where advance of recession had taken place during the past two centuries.

It was found that coastline erosion had been more extensive than accretion. Natural progradation during the past two centuries has taken place only on limited sectors, notably on parts of developing spits and cuspate forelands, as on Grenen (Fig. 19), the shifting spit that forms the northernmost point of Jutland, and on the Flahket, a sandy cuspate foreland that began to develop on the north coast of the island of Anholt after 1899. There has also been deposition within certain embayments, such as Køge Bay and Sjerø Bay in Zealand, and on the other barrier island shores of Fanø, Mandø, and

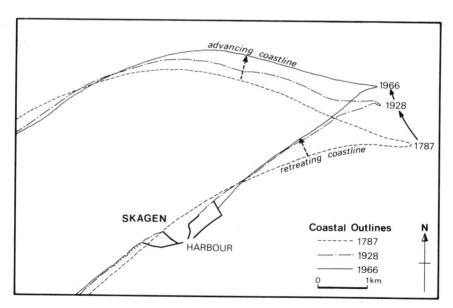

Fig. 19 Changing alignments of the cuspate spit of Grenen, at the northern end of Jutland, Denmark, from 1787 to 1966. Erosion of the southern flank (complicated by the effects of harbour construction at Skagen) has been accompanied by accretion on the northern flank (Bird, 1974)

especially Rømø, in south-west Jutland. In addition there has been localized beach progradation alongside man-made structures such as harbour break-waters, and some parts of the coast have advanced as the result of embanking and land reclamation in areas that were previously subject to tidal submergence, notably at the south of Mariager Fiord and on the marshy Vadehavet coast of south-west Jutland, where seaward advance of salt marsh has continued (Møller, 1963). The rest of the Danish coastline has either remained in position, or has receded during the past two centuries, erosion having been most rapid and extensive along the sandy North Sea coast of Jutland. At Klim, repeated surveys on an 850 metre segment of sandy coastline between 1968 and 1979 indicated irregular beach recession, averaging 5 metres per year, the dune cliff to the rear being cut back intermittently, up to 40 metres in stormy episodes (Christiansen and Møller, 1980); some of the sand has moved alongshore, for the coast to the east has shown a compensating progradation.

Several factors have influenced this pattern of coastline change. Sediment yields from the small Danish rivers have been very limited, most of them discharging into the Limfjorden and other lagoons and estuaries rather than directly on to the coast, although in south-west Jutland the Varde and the several smaller streams draining into the Vadehavet have been delivering fine-grained sediment, including some sand, to the tidal environment. Otherwise, the chief sources of sediment for coastal deposition have been from receding cliffs and eroding beaches, as well as from the sea floor, especially where this consists of sandy glacial drift. On the North Sea coast the net sediment flow during this period has been away from the eroding coastline, except in the southernmost sector where sea floor sand has been moving on to the outer shores of Fanø and the other barrier islands.

A special case of local progradation has been documented at Hoed, on the Djursland peninsula of eastern Jutland (Fig. 20), where gravelly waste dumped from a coastal limestone quarry has been distributed by wave action to build a shingle foreland, with stages of growth marked by beach ridges (Bird and Christiansen, 1982). Alternations of advance (landslide lobes) and retreat (cliffing) have occurred on the clay bluffs on Røsnaes Peninsula, north Zealand, on Røjle Klint, in north-west Fünen, and on Helgenaes in East Jutland (Prior, 1975).

The long-term tilting of the Danish land area, part of the broader pattern of isostatic recovery following deglaciation in Scandinavia, is still continuing, the northern regions rising while those to the south subside. In northernmost Denmark the land is thought to be rising at just over 0.5 millimetres per year, but this has not prevented erosion on the north Jutland coast. The subsidence of southernmost Denmark, on the other hand, has failed to prevent continuing accretion on the barrier islands bordering the North Sea. It appears that the tilting has been too slow to have influenced modern patterns of erosion and accretion, but it has probably contributed to the emergence and progradation seen since 1786 around the shores of Laesø

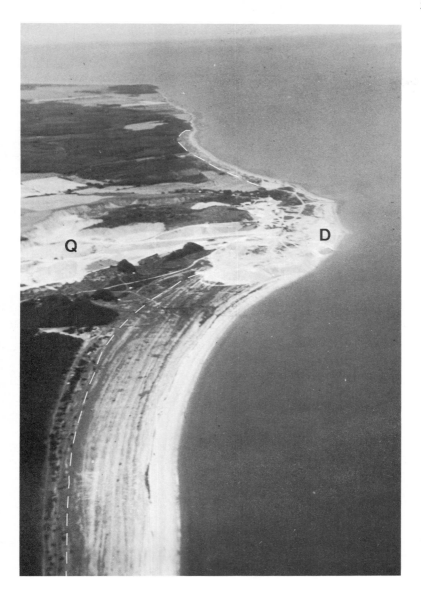

Fig. 20 The coastline at Hoed, Denmark, has been prograded by the accretion of parallel ridges of shingle derived from flint gravel waste, the unwanted residue of limestone quarrying at Q, dumped on the shore at D. The position of the coastline in 1784 is indicated by the pecked line. In the foreground the coastline has advanced 250 metres since that date.

Photo: Eric Bird (June 1974)

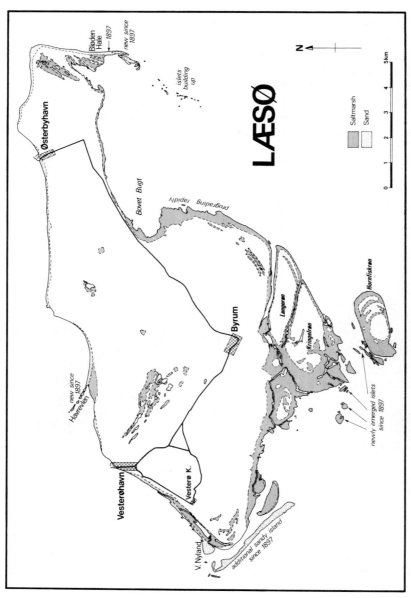

Fig. 21 The Danish island of Læsø, in the Kattegat, has increased in area since it was mapped in 1897. Continuing isostatic uplift has resulted in the emergence of shoals as new islands, and the enlargement of earlier islands and progradation of coasts by sandy beach accretion and the seaward spread of salt marshes

(Fig. 21), a Kattegat island which is relatively sheltered from strong wave action. Accretion on the island of Kyholm between 1933 and 1978 was correlated with widespread die-back of sea floor eel-grass (*Zostera*) which previously had prevented the shoreward drifting of sandy sediment to beaches around the island's harbour (Christiansen *et al.*, 1981).

The most striking changes are found along the North Sea coast of Jutland, where earlier sand deposition produced extensive dune systems which are now being trimmed back on their seaward margins (Fig. 22). The underlying glacial drift is exposed in receding cliffs at Hirtshals, Rubjerg and Bovbjerg, and the chalk basement appears in the cliff at Bulbjerg. Cliff recession here isolated a chalk stack, which was gradually reduced by marine erosion, and eventually collapsed in 1975. The sandy barriers which enclose the western arm of the Limfjorden, Nissum Fjord and Ringköbing Ford have all been cut back by erosion, the Limfjorden barrier beach coastline having retreated by up to 1500 metres since 1790. However, there has been local accretion adjacent to harbour breakwaters at Hanstholm, Hirtshals, Torsminde and Hvide Sande, representing a limited readvance on sectors that had previously been receding (Fig. 23).

Fig. 22 Concrete blockhouses built by the German occupying forces to defend the west coast of Denmark in the early 1940s have been undermined by subsequent recession of the sandy coastline. Here at Vigsø the coastline retreated about 150 metres between 1944 and 1974. Photo: Eric Bird (July 1974)

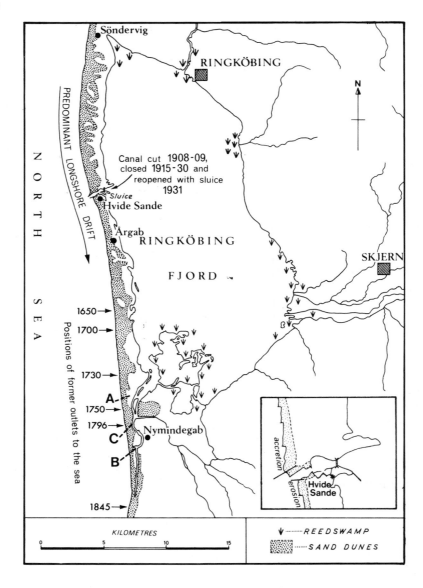

Fig. 23 The entrance through the sandy coastal barrier that encloses Ringköbing Fjord, a lagoon on the west coast of Denmark, migrated southwards as the result of longshore drifting of sand from the north. By 1845 it had been deflected far to the south, and artificial cuts were made at A (1845), B (1892) and C (1905) each of which soon became choked with sand. In 1908–9, a permanent artificial entrance, bordered by break-waters, was established at Hvide Sande (inset). Since then, sand drifting south has accreted alongside the northern breakwater, while the shore to the south, deprived of this sand, has been cut back. Based on data supplied by J. T. Møller

WEST GERMANY (North Sea Coast)

The North and East Frisian Islands have changed in configuration, some parts of the coastline being cut back while others have prograded or grown out as spits. Mroczek (1980) presented a map (1 : 200,000) showing gains and losses since the 1878 surveys of Schleswig-Holstein and Neuwerk/Schar-hörn and the 1891 surveys of Lower Saxony. On the west coast of Sylt, the Rotes Kliff, cut in glacial drift, has receded at up to 30 metres per year, but sand drifting northward and southward has been accumulating in sandy forelands at Ellenbogen to the north and Hörnum Odde to the south, thus lengthening the island (Petersen, 1978). On the East Frisian Islands, the North Sea sandy coastlines have retreated, while some accretion has proceeded on spits, especially at the eastern ends. Some islands have disappeared (e.g. Buise, near Nordeney); others are new (e.g. Mellum, a sand island that appeared in 1870 and is now 0.64 square kilometres with dunes up to 40 metres high) (Luck, 1978). Salt marshes have become more extensive locally on the inner (Wadden Sea) shores, and on either side of the Hindenburg Dam, built in 1928 to attach Sylt to the mainland (Gierloff-Emden, 1985b). Extensive areas have been artificially embanked and reclaimed, especially on the mainland coast, as at Wilhelmshaven. Historically, this coast has been submerged and re-shaped during successive North Sea storm surges, the Marcellus Flood of 1362 resulting in a sustained advance of the sea on the coast of Schleswig Holstein. The more recent storm surges (1953, 1962, 1973, 1976) have modified coastal features, but the drained and reclaimed areas have been maintained (Rohde, 1978).

Historical information on Heligoland dates from the writings of Adam von Bremen (c. 1050), and maps drawn in 1570, 1600, 1639, 1643, 1719, 1753, 1787, and at frequent intervals subsequently have been used to trace the reduction of this island by erosion. It now consists of two islands, one low and sandy, the other high, with sandstone cliffs that have receded historically, but are now artificially stabilized (Kramer, 1978).

NETHERLANDS

The West Frisian Islands have also shown changes in configuration over the past century: in general their North Sea coasts have been cut back, and irregular growth has occurred laterally, alongside the intervening *zeegats* (tidal entrances). Continuing submergence, due to a combination of rising sea level and land subsidence, explains the prevalence of erosion on these island coastlines, but marshland encroachment, complicated by artificial reclamation, has been extensive around the shores of the Wadden Sea, in front of dykes. South from Den Helder the coastline resembles that of western Denmark, with the dune fringe being cut back by marine erosion, and some stretches held by sea walls. Measurements by Edelman (1977) showed recession between 1860 and 1960 south to Umuiden, where accretion

58

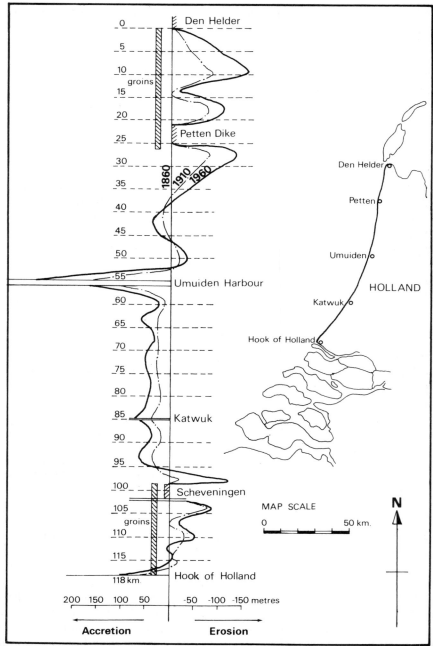

Fig. 24 The pattern of advance and retreat on the Netherlands coast, surveyed in relation to the 1860 alignment (taken conventionally as a straight line), shows gains and losses 50 and 100 years later. The coastline has been measured at the dune front, equivalent to high spring tide level (Edelman, 1977)

of up to 300 metres had taken place on either side of harbour breakwaters, progradation of up to 80 metres between Umuiden and Scheveningen, and erosion in the sector south to Hook of Holland (Fig. 24). There have also been changes around the deltaic islands at the mouths of the Rhine as a result of continued deposition of sediment and artificial reclamation, culminating with dam construction across the river mouths, leaving only the southernmost open to give access to the harbour of Antwerp.

BELGIUM

The 65 kilometre coastline of Belgium is sandy and backed by dunes. A. Ozer reported that west of Wenduine the beaches were broad, very gently shelving, up to 500 metres wide at low tide, protected from strong wave attack by offshore bars in shallow seas, and thought to have been stable or prograding during the past century. East of Wenduine the beaches are narrower and steeper, and not so protected; there has been beach erosion and coastline retreat, partly modified by the construction of sea walls and groynes and by artificial beach nourishment, but at large breakwaters (as at Zeebrugge) there has been no sustained accretion (Depuydt, 1972).

BRITISH ISLES

Reference has been made (pages 8–10) to the assessment of the question of coastline changes around the British Isles early in the present century by the Royal Commission on Coast Erosion and Afforestation (1907–11). In their final report, the Commissioners concluded that

far more land has been gained by accretion and artificial reclamation in recent years than has been lost by erosion. The gain has been almost entirely in the tidal estuaries. . . . The loss has been chiefly on the 'open coasts'. Had attention been confined to 'open coasts', exposed directly to wave action from the Atlantic Ocean, the North and Irish seas, and the English Channel, the land lost by erosion would certainly have exceeded the land gained by accretion or reclamation. On the other hand, the low-lying coastline around The Wash has been substantially advanced by embanking and reclamation.

Subsequent studies of the British coastline (many of which are summarized in Steers, 1964, 1973) have shown these general conclusions to be valid for the period up to the mid-1970s (Bird and May, 1976). Over the past century, trends of coastline change can be documented with reference to successive editions of Ordnance Survey maps and more recent changes can be traced on sequential air photographs (Fig. 25). Changes have been limited and localized on the hard rocky coasts of western Britain, even on sectors exposed to high wave energy from the Atlantic Ocean, but locally in western Ireland there are receding cliffs in glacial drift, and areas where such features as drumlins have been eroded, yielding material for the growth of related spits and tombolos (Guilcher, 1965). On the eastern and southern coasts of

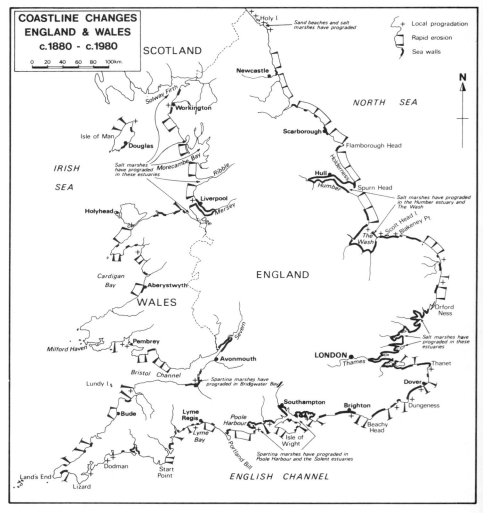

Fig. 25 Comparison of late nineteenth century maps with the 1980 outline of England and Wales has identified the pattern of coastline changes shown here. Revised from May (1979)

Britain, and around the Irish Sea, there are many places where cliffs in relatively soft sedimentary rocks, including glacial drift, have receded more than 50 metres during the past century, some of the most rapid recession being on the boulder clay cliffs of the Holderness coast, south from Bridlington in Yorkshire (Fig. 26). Here Valentin (1954) measured some 307 cross-sections on alignments where the 1952 configuration could be compared with that of 1852, and found that the cliffs had receded up to 200 metres (on average, 120 metres) in the intervening century. Nothing is known of sea

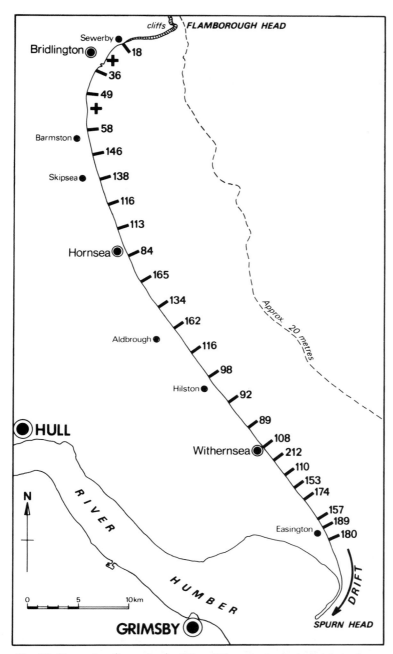

Fig. 26 Recession (in metres) of the cliff crest at selected points along the Holderness coastline in eastern England between 1852 and 1952 (from Valentin, 1954). The + symbols indicate two sectors where local progradation occurred during this period

floor changes during this period, but it is possible that the sea floor, also cut in boulder clay, has been lowered by erosion in such a way as to maintain the transverse profile of the coast, which has thus migrated landward in response to the energy input of wave action from the North Sea. In general, cliffs cut in soft rock, within embayments flanked by protective headlands (e.g. Filey Bay in Yorkshire), have retreated more slowly than those on more exposed coastlines like Holderness. De Boer (1978) pointed out variations in the rate of cliff retreat in Holderness related to minor structural features; the exposure of a protective ramp of boulder clay on the foreshore can reduce the erosion rate on cliffs of softer material to the rear. Pringle (1981) has shown that cliff recession accelerates with the passage of low, narrow beach sectors (*ords*) between migrating beach lobes, which are higher and relatively protective, on the Holderness coastline. The arrival of southward-drifting beach material on Spurn Head has contributed to the recurrent growth of the spit, as described by De Boer (1969), in relation to the historical recession of the cliffs of southern Holderness over recent centuries (Fig. 27). The spit attained its maximum extent on successively set-back alignments in

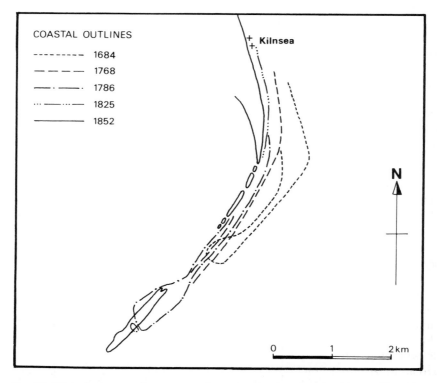

Fig. 27 Historical maps show successive stages in the growth of Spurn Head at the southern end of the Holderness coast (cf. Fig. 26) on set-back alignments related to recession of the Kilnsea cliffs (after De Boer, 1969)

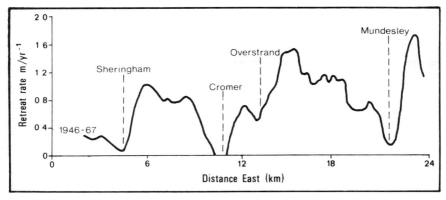

Fig. 28 Graph showing varied rates of cliff retreat between 1947 and 1967 on the North Norfolk coast (after Cambers, 1976)

c. 1350, c. 1600 and c. 1850, with intervening phases of dissection and landward migration (Steers, 1964). In recent years the building of groynes has reduced the longshore sediment supply from the north, and sea walls have given Spurn Head an increasingly artificial outline.

Off the North Norfolk coast between Sheringham and Cromer the sea floor consists of resistant chalk which underlies lhe glacial drift of the receding cliffs, and here it is likely that cliff recession will gradually wane as the shallow offshore zone widens and wave energy consequently diminishes. Lateral variations in the rates of cliff recession on the North Norfolk coast between 1946 and 1967 are shown in Fig. 28. Recurrent landslides have led to cliff-crest recession and alternations of coastline advance (up to 60 metres near Cromer in 1962) and retreat as landslide lobes develop, and are consumed by marine erosion (Hutchinson, 1976) (Fig. 29). Cambers (1976)

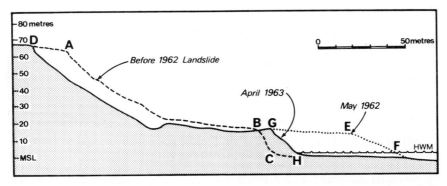

Fig. 29 Changes in the transverse profile of the coast near Cromer, North Norfolk, England, as the result of a 1962 landslide. The previous profile (ABCH) showed cliff-crest recession from A to D and basal lobe formation (GEF) immediately after the landslide, but by April 1963 the lobe had been cut back to GH by marine erosion (after Hutchinson, 1976)

suggested that cliff erosion on the east coast of England (Suffolk, Norfolk, Holderness) could be interpreted as a dynamic equilibrium (i.e. a condition whereby the forms and processes are recurrently in a steady state of balance) over periods of the order of a century, implying that after such an interval they would have returned to the same form, the cliff crest and cliff base having both retreated about 100 metres. Certainly erosion has been long-continued, and historical records have been used to trace the disappearance of coastal settlements such as Dunwich (Parker, 1978) as the result of cliff retreat over several centuries.

Chalk coasts in south-eastern England have shown intermittent recession (May, 1971) as the cumulative result of many localized and occasional rock falls (Hutchinson, 1971): these are particularly frequent after cold winters when the chalk is loosened by the expansion of freezing groundwater, and then disintegrates during the thaw. Recession of cliffs cut in soft Tertiary clays has been rapid on the shores of Barton Bay, Hampshire (Fig. 30), and on the Dorset coast there have been cyclic changes due to landsliding and cliff-base recession in Jurassic and Cretaceous clays, sands and limestones (Figs. 31–33). At Fairy Dell, near Charmouth (Fig. 34), the upper and lower cliffs both retreated about 36 metres between 1887 and 1968, the intervening zone showing complex slumping (Brunsden and Jones, 1980) (Fig. 35). On the resistant Lower Palaeozoic rocks of south-west England, cliff retreat during the 5000 years since the Holocene marine transgression established approximately present sea level has been more limited; coasts sheltered from strong wave action show slope-over-wall profiles which retain extensive relics

Fig. 30 Slumping clay cliffs on a receding coastline at Barton Bay, southern England. Photo: Eric Bird (March 1962)

Fig. 31 Receding cliffs on the coast of Lyme Bay, Dorset, England. In the foreground a sea wall has been built to halt cliff erosion. Maintenance of natural coastal scenery on such a coastline is aided by the presence of a broad beach, which reduces the frequency and vigour of wave attack on the cliff base. Photo: Eric Bird (November 1977)

Fig. 32 Landslide lobe below Golden Cap, Dorset, resulting in local progradation of the coastline, with large boulders pushed out towards the seaward margin. Photo: Eric Bird (November 1981)

Fig. 33 Festoon of boulders in the sea beneath Golden Cap, Dorset, indicating the former margins of a landslide (cf. Fig. 32), the softer clay having been washed away by wave action. Photo: Eric Bird (November 1981)

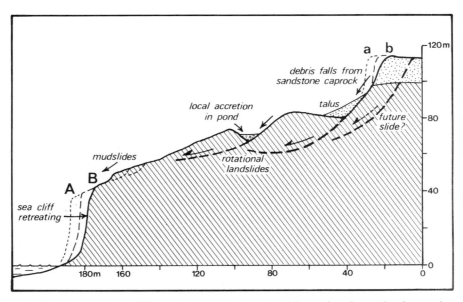

Fig. 35 Evolution of a cliff subject to recurrent landslides and undercutting by marine erosion near Lyme Regis, southern England. During the past century cliff crest recession (a to b) has been roughly the same as undercliff recession (A to B), with complex variations in the intervening landslide topography (after Brunsden and Jones, 1980)

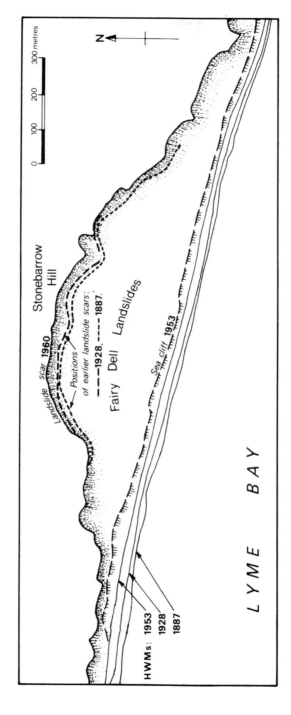

Fig. 34 Retreat of the cliff-crest (landslide scar) and cliff base (high water mark) between 1887 and 1960 on the Dorset coast of southern England east of Lyme Regis (Brunsden and Jones, 1980)

of a degraded coastal slope, mantled with periglacial drift, inherited from a colder phase in late Pleistocene times, with only the lower part actively cliffed. Recession of cliffed coastlines has locally been accelerated by quarrying, for example on the Purbeck peninsula, at Lyme Regis, and at Llantwit Major in Wales (Williams and Davies, 1980), or by the removal of beach material, as at Hallsands in south Devon.

Sectors of natural beach progradation resulting from accretion of coastal sediment have been localized. In some places accretion has simply resulted from the arrival of sediment eroded from an adjacent sector and carried alongshore. Where the eroded coast is low-lying the area of depositional land gained may be similar to the area of erosional loss. Examples of such lateral transference of coastal sediment include Winterton Ness and Benacre Ness on the east coast of England, Dungeness on the south coast, the paired spits in Bideford Bay and Carmarthen Bay, the forelands of Morfa Dyffryn and Morfa Harlech, and the coastline of south Lancashire between Formby and Southport. On the North Devon coast at Northam, Steers (1964) has illustrated the recession of the southern part of a shingle beach between the mid-nineteenth and mid-twentieth century, accompanied by beach ridge accretion to the north, changing orientation by about 10 degrees. At Dungeness, long-continued erosion of the southern flank has supplied longshore drifting around the point to prograde the eastern shore (Steers, 1964).

Another kind of progradation has occurred where spits have grown parallel to the coast, as at Orford Ness, where such lateral growth has effectively advanced the coastline seaward by at least a kilometre. Carr (1969) traced historical changes here, especially fluctuations at the distal (southern) end between 1804 and 1967. At Blakeney Point the coastline has retreated as the result of recurrent storm surges driving the main shingle bank southwards (35 metres in the 1953 surge) (Figs 36, 37), but there have been intricate changes on Far Point at the western end, where new recurved hooks have been added during the past century (Fig. 38). Similar changes took place on Scolt Head Island, to the west between 1891 and 1958 (Steers, 1960). The pattern of erosion and deposition has in some places been related to wave refraction across deeps and shoals in nearshore waters, and changes in such sea floor topography can lead to changes on the adjacent coastline. The recent evolution of Winterton Ness and Benacre Ness have been related to nearshore changes of this kind. Erosion of the flanks of Magilligan Foreland, Northern Ireland, at rates of up to 6 metres per year, and rapid progradation at its north-western apex, are changes resulting from the movement of nearshore shoals (Carter, 1979).

On the North Sea coast, substantial changes have occurred during successive storm surges, particularly those of 1953 and 1978, when many beaches were scoured away, and cliffs in soft sedimentary rock were cut back several metres in a few hours (Steers *et al.*, 1979). In general the beaches were restored in succeeding calmer weather. Measurements of volumes gained and lost on such a coastline can be assembled in the form of sediment budgets

Fig. 36 The beach at Blakeney, Norfolk, retreated several metres during the January 1978 storm surge, when the shingle ridge was driven inland over salt marsh (cf. Fig. 37). Peaty clay outcrops in the beach face, marking the level of the overrun salt marsh. Photo: Eric Bird (July 1979)

Fig. 37 Fans of shingle washed on to the salt marsh behind the beach at Blakeney, Norfolk, during the North Sea storm surge in January 1978, when waves overtopped the beach (cf. Fig. 36). Photo: Eric Bird (July 1979)

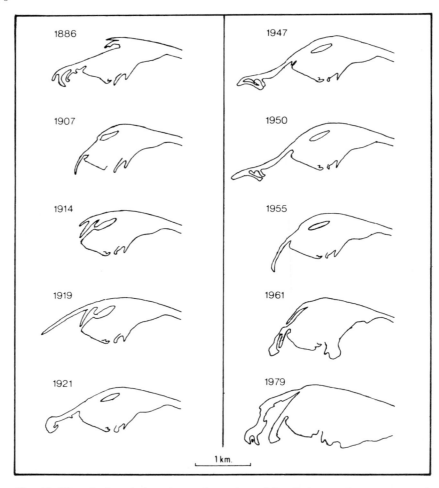

Fig. 38 Historical variations in configuration of Far Point, at the western end of Blakeney Point, Norfolk, England, from 1886 to 1979, traced from successive maps and air photographs (supplied by Juliet Bird)

which quantify onshore, offshore and alongshore transportation. Sediment budgets for the East Anglian coast, where there has been a net loss of sediment from the coastline to the sea floor during the present century, have been estimated by Clayton (1980), and are shown in Fig. 39.

Coastal accretion may also result from the delivery of fluvial sediment, or from shoreward drifting of sediment derived from the sea floor, but examples of this have been exceptional around the British Isles during the past century. The types of sediment supplied by the British rivers vary considerably, but few rivers carry sand or gravel to the coast in substantial quantities, and there are no true deltas. Rivers draining areas of former glacifluvial deposi-

tion, such as the Nairn and the Spey in north-east Scotland, have been
delivering sand and gravel to the coast for incorporation in beaches near
their mouths. Rivers that drain steep catchments, such as the Lyn in north
Devon, carry coarse debris to the coast during floods, and have built small
deltaic fans, mainly inter-tidal.

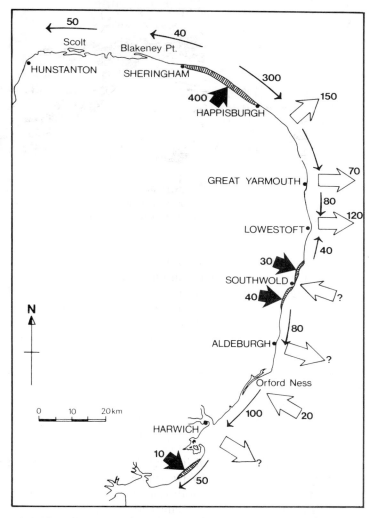

Fig. 39 Sediment budget for the East Anglian coast (Clayton,
1980). The coastal system has gained sediment from cliff erosion
(black arrows) and onshore drifting from the sea floor (at
Southwold and Orford Ness). This has been moved alongshore to
accreting sectors such as Blakeney Point and Scolt, and there have
been losses to the sea floor as indicated. Estimated volumes in
thousands of cubic metres per year

In Cornwall, streams draining mining regions have delivered sand and gravel to the coast at Par and Pentewan near St Austell (Everard, 1962), and in St Ives Bay, during recent decades. This yield is largely due to disturbance of the land surface within their catchments and along the valley floors by mining operations. Most English rivers carry only silt and clay to the sea, any coarser sediment being relinquished within estuaries, which are essentially valley mouths drowned by the Holocene marine transgression and not yet infilled by sedimentation.

There is no doubt that sand and gravel derived from the sea floor have been delivered to sectors of the British Isles coast during and since the major Holocene marine transgression. Shingle accumulations such as Chesil Beach and coastal sand masses of the type seen in north Cornwall and South Wales probably originated in this way, but the extent to which such material is still coming ashore is difficult to judge. Chesil Beach (Fig. 40) is being driven intermittently landward, so that the coastline is retreating, but the rate of retreat during the past century has been slow (Carr and Gleason, 1972). On the South Wales coast the sandy beach and dunes at Pendine have been cut back in recent decades, but accretion has continued at Pembrey with the addition of new grassy dunes on the prograding coastline. At Tentsmuir Point on the east coast of Scotland (Steers, 1973), the sandy beach prograded by up to 680 metres between 1836 and 1965, its more recent advance being measurable with reference to a line of concrete anti-tank traps built on the backshore in 1940 and now up to 330 metres inland. Continuing progradation here may be explained by sea floor derivation, as could the sand accumulation in progress on the north of Rattray Head, on the Sands of Forvie, in the Holy Island area, and along the north Norfolk coast. In some cases, sand washed onshore to beaches has been swept on by wind action to accumulate as backshore dunes, so that despite accretion the coastline has not prograded. However, Blakeney Point in Norfolk is still receiving some material, especially gravel, as a consequence of cliff erosion to the east, and part of the recent accretion in this area could be due to the redistribution of sediment already present in the inter-tidal zone along the Norfolk coast, rather than to shoreward supply. On the Atlantic coasts of Britain there is little evidence for a continuing sand supply from the sea floor, except for the shelly material still arriving on some of the 'machair' beaches in north-west Ireland and the Outer Hebrides (Ritchie, 1985).

Evidence from tide gauge records around the British Isles suggests a rising of the land in the north and a subsidence in the south relative to mean sea level. According to Valentin (1953) the most rapid uplift of the coast (3 to 4 millimetres per year) is in progress towards the heads of Scottish firths and lochs, and it is possible that sand accretion in north-east Scotland has been facilitated by this continuing emergence. Similarly, the subsidence in progress in East Anglia, around the Thames estuary and along the south coast of England (1 to 2 millimetres per year) may have contributed to the prevalence of erosion on these coasts.

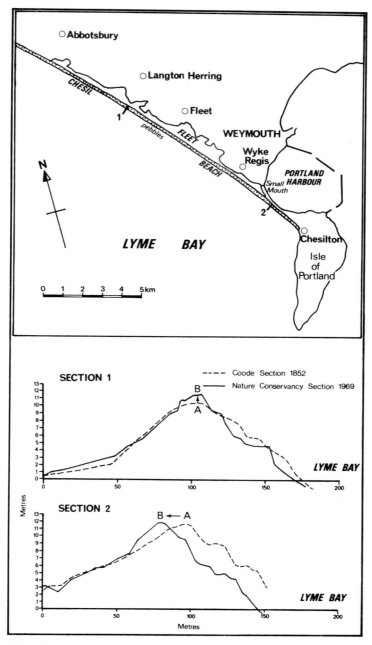

Fig. 40 Comparison of profiles across Chesil Beach, surveyed by Sir John Coode in 1852, and Carr and Gleason in 1969. At Section 1 the beach crest has been built up without any landward displacement, but at section 2 the beach has been driven landward. These changes probably took place during storm surges which overtopped the beach, as in 1954 (after Carr and Gleason, 1972)

The seaward spread of salt marshes during the past century has been extensive in the Solent region, in Poole Harbour, at Bridgwater Bay, and in a number of estuaries around Britain, especially since the advent of *Spartina anglica*, thought to have originated in Southampton Water around 1870. In the Dee estuary, for example, *Spartina* introduced in 1922 has formed wide marshland terraces where previously there were unvegetated mudflats (Marker, 1967). In some areas, however, 'die-back' in recent decades has led to the modification, and in places the reduction, of *Spartina* marshes, and it is common to find an erosional cliff marking the seaward limit of salt marshes. Recession of seaward margins of salt marshes has been widespread in southern England, where it may reflect a modern sea level rise indicated by historical changes on tide gauges, as at Newlyn, in Cornwall, where the level of mean tides rose 12 centimetres in the half-century following determination of Ordnance Datum here in 1915–21 (see p. 170). On the other hand, marshes have spread seawards in estuaries where mining activities within the catchment have increased fluvial sediment yields, as in the Fal estuary at Ruan Lanihorne (Fig. 41) (Ranwell, 1974).

Examples of man-induced coastline progradation are found where groynes or breakwaters have intercepted the longshore drift of beach material to form local depositional areas. On the south coast of England such accretion is common on the western side of protruding structures as at Newhaven Harbour, the dominant drift of beach material being from west to east, while on the east coast it is the prevailing southward drift that has been intercepted at such places as Great Yarmouth. Man's impact is also seen where waste gravel dumped from coastal quarries has prograded beaches in coves on the east coast of the Lizard Peninsula in Cornwall, notably at Porthoustock and Porthallow, while the dumping of coal mining waste and basic slag from a steel works has led to beach accretion on the Cumberland coast. On the other hand the removal of sand or gravel from beaches may have accelerated coastal erosion by allowing stronger wave action to reach the backshore during storms and high tides. At West Bay in Dorset, where gravel has been extracted from the beach over several decades, cliff erosion has become severe and it has been necessary to build and elaborate coastal defence structures to prevent further land losses. Accelerated cliff erosion has followed the quarrying of beach gravel at Gunwalloe, near Helston in Cornwall. Beaches that protect coastal slopes in soft, erodible material should not be depleted by such quarrying; indeed, it would be preferable to maintain them as natural protective formations by dumping additional sand or gravel on the shore.

N. Stephens reported that on the 2034 kilometres coastline of Ireland only 30 kilometres had advanced and 231 kilometres had retreated, the remaining 1773 kilometres having remained stable over the past century. On the hard rock cliffs of western Ireland changes had been very slow, but there have been sporadic rock falls and slides. On the Aran Islands, R. W. G. Carter found that cliff recession had cut across the Iron Age hill fort at Dun

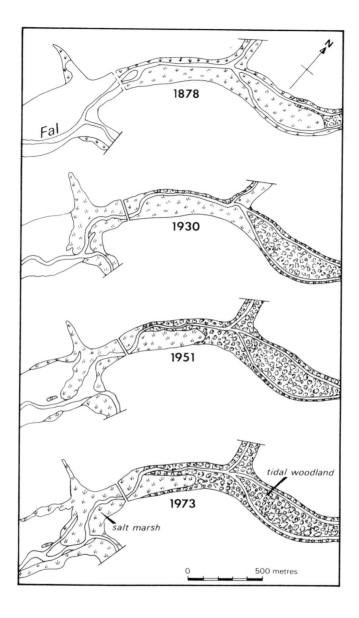

Fig. 41 Historical changes in the extent of salt marsh and tidal woodland (mainly alder and willow scrub) in the Fal estuary, south-west England, from 1878 to 1973. The advance of the salt marsh followed sedimentation, accelerated here by the arrival of clay washed down from kaolin quarries in the headwaters of the Fal (after Ranwell, 1974)

Aengus, indicating an annual cliff retreat of about 0.01 metres over the past 2000 years. On the east coast of Ireland the intermittency of cliff recession (up to 1.1 metres per year) in glacial sands and gravels on the coast of County Wexford may be related to the migration of subaqueous banks parallel to the coast, 2 to 5 kilometres offshore, erosion being accelerated when larger waves arrive through deeper inshore water between these banks.

In south-east Ireland construction of a harbour at Rosslare may have increased erosion of the cliffs to the north. Gravelly barriers on the south coast of County Wexford are transgressing (being driven landward), the coastline retreating up to 2 metres per year, partly because of commercial gravel extraction (Orford and Carter, 1982).

At Port Stewart (Derry) and Portrush (Antrim) dunes have been cut back at about 20 centimetres per year, with occasional storm erosion of more than 4 metres, but much of the sediment returns to the beach and dune fringe within 40 to 80 days of such storms (Carter, 1980). Near Port Ballintrae, County Antrim, a small beach disappeared after a jetty was constructed, and beach reduction has resulted in the acceleration of cliff retreat to more than 0.3 metres per year (Carter *et al.*, 1983).

Further details of coastline changes in the British Isles during the past century were given in Bird and May (1976). The prevalence of beach erosion is indicated by the fact that dunes behind sandy beaches typically show cliffed seaward margins.

FRANCE

The north coast of France, like the south coast of England, is predominantly cliffed and receding, especially on the chalk outcrops near Calais and between Ault and Le Havre. Prêcheur (1960) reviewed historical records and estimates of recession of chalk cliffs, and found they were retreating at between 0.08 and over 0.80 metres per year, the most rapid change being at Bourg d'Ault, which lost over 97.5 metres between 1825 and 1955. As in southern England, freezing and thawing in cold winters (e.g. 1962–3) have resulted in disintegration and rock falls. Cliff recession near the Cap d'Ailly lighthouse on the Picardy coast amounted to 110 metres in 70 years (Ottmann, 1965). Briquet (1930) traced changes on the low-lying coasts of Flanders and Picardy with reference to historical maps. The Flanders coast is now largely artificial and changes have been slight, but the spits bordering the estuaries of the Canche, the Authie, and the Somme have changed in outline during the past century, with growth on the Pointe du Touquet, at the Pointe de Pouthiauville and the Pointe du Hourdel, each nourished by longshore drifting from the south. At Berck Plage there has been substantial recession since De La Favolière's survey in 1671 (Briquet, 1930).

The older and harder rocky formations of the Cherbourg Peninsula and the Brittany coast have shown little change, but beaches of sand and shingle have been modified in recent decades. Growth of the compound recurved spit of Penn an C'hleuz on the north Brittany coast from 1823 to 1976 has been traced by Hallegouët and Moign (1976); its growth has been matched by the driving back and destruction of the spit of Peleuz, on the opposite shore (Fig. 42). A. Guilcher reports that since 1978 progradation has given place to recession on the shore of Penn an C'hleuz, evidently because of removal of calcareous sediment from the foreshore by local farmers to lime their fields. Changes have also occurred in the salt marshes bordering the macrotidal Bay of St Malo and its tributary estuaries, where Verger (1968) recorded short-term advances and retreats of the seaward margins: recession of up to 2 kilometres between 1903 and 1924, advance of up to 1 kilometre from 1924 to 1945, then renewed retreat. Salt marshes in the Anse de l'Aiguillon near La Rochelle advanced up to 130 metres between 1934 and 1955, while those at the mouth of the Somme increased in area by 13.5 hectares per year from 1971 to 1975.

On the Atlantic coast there has been continued accretion on portions of sand spits, notably the Pointe d'Arcay in Vendée where Verger (1968) traced the addition of further recurves between 1923 and 1967 from air photography,

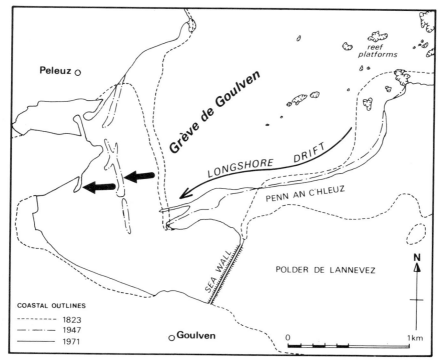

Fig. 42 The growth of the spit at Penn an C'hleuz, in northern Brittany, France, from 1823 to 1971, was accompanied by the retreat and dissection of an earlier spit on the western side of the Grève de Goulven (after Hallegouët and Moign, 1976)

and the Pointe de la Coubre, which has continued to grow southwards into the Gironde estuary (Facon, 1965). Guilcher (1985) reported evidence here of an advance of 3.4 kilometres southward between 1825 and 1971, part of the accretion being derived from erosion of the western shore, facing the Bay of Biscay, and part from a large nearshore submarine sand bank. Tidal scour in the Gironde estuary is now impeding further southward extension, the spit having grown round into the bay of Bonne Anse. Farther south, the long sandy coastline of the Landes region has not been prograding. A. Guilcher reported long-term retreat of up to 10 kilometres since the sixth century AD, accelerating in the last two centuries: German bunkers built on the backshore dunes in the 1940s have been undermined, and are now at the low tide line. By contrast, there have been minor additions to the southern end of the Cap Ferret spit at the entrance to the Bassin d'Arcachon in recent decades. The present straight sandy coastline between Pointe de Grave and the mouth of the Adour River near Bayonne has been retreating by up to 10 metres per year since 1881, the sand being carried southwards by the obliquely arriving north-westerly swell.

SPAIN AND PORTUGAL

The north coast of the Iberian peninsula is generally a hard rock coast, often steep, and intersected by rias, within which deposition is proceeding (Hernandez Pacheca, 1966). No examples of continuing progradation have been reported from here, but I. Asensio Amor has described short-term cyclic variations on beaches of the Gulf of La Masma (Lugo) and traced the growth of sandbanks since 1864 within the Ria del Eo. Reclamation has reduced the area of Santander Bay by 46% since 1837.

On the Portuguese coast there are long sandy beaches, some of which enclose lagoons with variable or artificially stabilized tidal entrances, while others end in spits bordering estuaries, longshore drift being generally southward. Accretion has taken place on the northern side of the breakwaters at Figueira da Foz, at Peniche, and at Vilamoura. It appears that the Atlantic coastline has been gradually receding over the past century, some of the sand having been added to spits at the inlets to give localized advance.

Sand accretion has continued on parts of the Algarve coast, and sand drifting eastwards is still being supplied to the coastline of the Gulf of Cádiz. Comparison of charts made in 1873 with more recent (1974) outlines (provided by M. Segado, Instituto Hidrografico de la Marina, Cádiz) showed eastward growth of the sand spit at Punta del Gato, in front of El Rompido on the Las Piedras estuary, and broadening of the Punta Umbria spit at the mouth of the Odiel River. Divergence of sand drifting from the receding central coast of Lower Andalusia has led to progradation north of Mazagon, and south of Matalascanas to the mouth of the Guadilquivir River (Vanney et al., 1979). On the Mediterranean coast of Spain there are extensive beaches, but evidence of continued accretion is limited to portions of the Ebro delta, and the mouths of smaller rivers such as the Fluvia, draining into the Gulf of Rosas, Catalonia. The coastline of the Llobregat delta retreated at 10 metres per year between 1950 and 1968 (Marques and Julia, 1983). Among many examples of localized accretion related to the trapping of longshore drift by artificial structures are the beach on the south side of the tombolo at Peñiscola, and the sand accretion on either side of the breakwaters at Vilanova i la Geltrú (Marques and Julia, 1983).

MEDITERRANEAN FRANCE

In southern France, the long low sandy coastline that extends from near Perpignan around to Agde is not advancing, but the margins of the Rhône delta have shown patterns of advance and retreat associated with variations in the size and sediment yield of the Rhône distributaries and with the dominance of longshore drift from east to west. Van Straaten (1959) traced the advance of the coastline on either side of the mouth of the Grand Rhône by 1300 metres between 1906 and 1954, and the growth of the spit of Pointe de la Gracieuse to the east. Pointe de Beauduc has grown out westwards following the decay of the Vieux Rhône after its abandonment in 1711 and the erosion of its former lobate subdelta, which has been cut back about 5 kilometres in 250 years at Faraman. Over a similar period, Pointe de l'Espiguette has developed west of the truncated subdelta of the Petit Rhône. Guilcher (1985) reported on the effects of barrage construction on the Rhône and its tributaries, which have reduced sediment yield from 40 million tonnes per year at the end of the nineteenth century to about 12 million tonnes in 1956–57, and only 4 or 5 million tonnes in 1970. In consequence, the deltaic coastline alongside the mouth of the Grand Rhône has failed to sustain the advance described by Van Straaten, and has been cut back 400 metres between 1954 and 1971. The Pointe de Gracieuse to the east has continued to grow, but it now receives material derived from deltaic coastline erosion rather than from direct accessions of fluvial sediment.

East of the Rhône delta the coastline is generally steep and rocky, and there has been little recent change in configuration either of cliffed sectors or of the sand and shingle beaches that line parts of the shore. The shingle is derived from fluvial gravels supplied by such rivers as the Var, which have steep courses draining catchments where glacifluvial deposits are extensively available. Onshore movement of sand to beaches in Provence in recent years has been correlated with the deterioration of nearshore sea floor *Posidonia* vegetation, which previously retained this sediment (Blanc and Grissac, 1978). This deterioration appears to have been due largely to man's impact: protection and rehabilitation of seagrass meadows has been successful in some areas (Meinesz *et al.*, 1978, 1981, 1983a, b; Thomas, 1983). At Marseilles, as elsewhere around the Mediterranean, archaeological sites indicate changes of sea level, due mainly to uplift or depression of coastal land, over the past 2000 years, but such changes have been minor during the past century.

CORSICA

Corsica has steep and rocky coasts, with some receding cliffs, notably on the limestones at Bonifacio, and an undulating east coast lowland, which is bordered by sandy beaches that locally enclose lagoons such as the Etang de Biguglia. There is little evidence of change during the past century, although it is probable that deltaic marshlands and associated beaches and spits at the mouths of rivers such as the Porto Vecchio, have been modified slightly over the period.

ITALY

Zunica (1976) prepared a report on coastal changes in Italy during the past century in which he compared coastal outlines shown on charts produced by the Uffico Idrografico della Marina and maps published by the Istituto Geografico Militare during the period 1863–92 with those of 1953–72.

Little change was detected on rocky shores, which make up a high proportion of the coasts of Sicily and Sardinia, and are also extensive in Liguria and parts of southern Italy. The greatest changes, both of advance and retreat, were along the Adriatic coast between the Po and the Tagliamento deltas (Zunica, 1971). Comparison of surveys dated 1887–92, 1908, 1931–8 and 1961–6 showed progradation on the shores of the Adige delta, and on breach sectors adjacent to breakwaters built alongside gaps through the sandy coastal barriers at entrances to the Venice Lagoon (Porto di Chioggia, Porto di Malamocco and Porto di Lido), and recession on the intervening segments, where protective sea walls have been built, as well as to the north-east, past Porto di Piave Vecchia (Zunica, 1971). Erosion has been accentuated by a rising sea level, due mainly to subsidence of the Venice region; the sea rose 0.22 metres here between 1908 and 1980. The mouth of the Piave has been deflected north-east by spit growth, and there have been minor gains and losses along the Caorle coastline, with progradation on the southern shore of the Tagliamento delta, especially west of Bibione (Zunica, 1971).

It is possible to trace changes on this coast over a long period with reference to Roman and Mediaeval descriptions: P. Fabbri reported that Ravenna, Adria and Aquilea were coastal or near-coastal towns in the Roman era, and over the succeeding centuries there has been overall progradation, leaving each of them several kilometres inland. Part of this is due to deposition of sediment from the Po. The Po delta grew substantially after its outlet was diverted (to prevent siltation of the Venice lagoon) in 1559–1604 to channel outflow eastward from Contarina, but in recent decades growth has slackened, partly because of subsidence. Changes along the coastline include the lateral growth of spits, the migration of beaches onshore over marshland, and the submergence of bordering lowlands (Zunica, 1976).

Elsewhere in Italy, sandy coastlines that have advanced are generally adjacent to river mouths, notably the Arno and Serchio which have nourished progradation of the Luni-Pisa sandy coastline (Mazzanti and Pasquinucci, 1983), the Ombrone, Tevere (Tiber), and Volturno on the west coast, and the Ofanto, Biferno and Tronto on the east. Long-term progradation has stranded ancient ports such as Pisa (Arno delta) and Ostia (Tiber delta) far inland. Substantial accretion has occurred on the shores of the Metaponto coastal plain, bordering the Gulf of Taranto, but in recent decades progradation has given place to erosion here, probably as a consequence of the development of irrigation schemes and the building of dams that impound reservoirs, reducing fluvial water and sediment yields. On the Latium coast, north and south of the Tiber delta, erosion has predominated, both on beaches and on sectors of cliff cut into soft rock formations, as at Anzio (Caputo *et al.*, 1983). Near the promontory of Arco Muto marine erosion is dissecting the ruins of Nero's villa, built over 1900 years ago against the cliffs (Fig. 43), and to the south coastline retreat has left the watchtowers of Torre Astura stranded offshore. Locally vulcanicity has prograded the coastline, for example around Stromboli and the Lipari islands. In Sicily the shores of the Gulf of Catania have prograded as the result of fluvial sediment supply, and there are several small bays in southern Italy, Sicily and Sardinia where short, steep streams have continued to nourish local beaches. On the Adriatic coast, Zunica suggested that progradation in the sector between Porto d'Ascoli and Giulianova was the outcome of 'coastal straightening', this being a sector of convergence of longshore drifting of material from eroding sectors to north and south.

There is evidence that many of the Italian sandy coastlines prograded during the eighteenth and nineteenth centuries, when deforestation and the intensification of cultivation and grazing in headwater regions increased fluvial sediment yields, leading to the delivery of larger quantities of sand and gravel to river mouths. On many sectors this progradation continued until the 1930s, when erosion began, and since the 1950s this erosion has become widespread and more severe. There has evidently been a reduction in the supply of sediment to these coasts. This could be the result of a climatic change toward drier conditions, but it is more likely that it results from man's activities, such as the building of reservoirs in river valleys for hydroelectricity generation and irrigation schemes, extraction of sand and gravel from river beds, and afforestation and soil conservation works in the hinterland. In addition, many coastal sectors have been urbanized and industrialized, and the building or extension of habour structures, breakwaters at river mouths and lagoon entrances, and sea walls designed to halt beach erosion have all resulted in modifications to natural shore dynamics, in many cases initiating or accentuating coastal erosion.

Fig. 43 The ruins of Nero's Villa, built in the 1st century AD, are being dissected by marine erosion on the coast near Anzio, Italy. It is evident that cliff recession has occurred, but the construction of the villa, the rear walls of which are set against the cliff, implies that there have also been changes of sea level, relative to the land, here during the past 2000 years. Photo: Eric Bird (February 1982)

Fig. 44 Beach erosion on the Adriatic coast north of Rimini has been halted by the construction of a chain of offshore breakwaters, formed of large limestone blocks, which greatly reduce incoming wave action. Photo: Eric Bird (May 1982)

Numerous offshore boulder breakwaters have been built parallel to the Adriatic coast north and south of Rimini in an attempt to reduce erosion and maintain recreational beaches (Fig. 44): the outcome is a sandy coastline with cusps in the lee of each breakwater (Cencini *et al.*, 1979).

MALTA

The coasts of the Maltese islands consist mainly of slowly retreating limestone cliffs, with some areas of collapsed boulders where underlying blue marls have been washed out. Beaches are limited to small embayments, such as St Paul's and Mellieħa Bays in northern Malta and Ir-Ramla in northern Gozo (Paskoff and Sanlaville, 1978).

YUGOSLAVIA

Steep limestone slopes dominate much of the Yugoslav coastline, with some actively receding cliffs on the Adriatic shores of outer islands such as Kornat. North-west of Split sand and gravel beaches are small and localized, usually near the mouths of river valleys. Alongside the Planinski Channel, for example, beaches fringe deltaic fans built by the rivers that drain the mountainous hinterland. At Starigrad such a depositional foreland is being cut back on the more exposed western flank, yielding beach gravels which drift alongshore to prograde the sheltered eastern flank. South-east of Split, sandy beaches, spits and tombolos are more extensive, incorporating sediment washed in from the sea floor as well as sand supplied by rivers, but historical changes have been slight on this coastline.

ALBANIA

Of the 385 kilometre Albanian coastline, 30% has been stable, 28% has advanced, and 42% has retreated during the past century (Shuisky, 1985a). Prograding sectors occur around the mouths of rivers, notably the Buna, Drin, Mat and Erzen Rivers in the north, and the Shkumbin, Seman and Viose in the central sector. Fluvial sand supplies have nourished deltas, spits, beach-ridge plains, and barriers that enclose lagoons. Progradation averaging 1.2 metres per year has stranded the ancient coastal town of Lesha 8 kilometres inland, but R. Paskoff reported rapid erosion of the sandy coastline of the old delta of the Seman River following its northward diversion. The southern coast of Albania is mainly steep, with some active cliff recession (up to 1.3 metres per year at Cape Durrësit) on limestones and sandstones. Beaches are supplied with cliff-derived sands and gravels, or fed with coarse loads from steep streams draining the coastal slopes.

GREECE

An alternation of steep, rocky coasts with beach-fringed embayments and inlets fed by rivers is typical of much of the Greek coast, and there has certainly been accelerated coastal deposition through historic times in several areas, notably near Salonika, where Fels (1944) documented enlargement of parts of the Axios delta into Thermaïkós Bay between 1831–4 and 1934, and within sheltered gulfs such as the Gulf of Amvrakia on the west coast. Kraft et al. (1977) traced longer-term changes in Thermaïkós Bay, where the ancient coastal town of Pella has been stranded several kilometres inland (Fig. 45) and in the Gulf of Messenia, much reduced by rapid growth of the Pamisós delta. In Maliakos Gulf, Tziavos (1978) showed that sediment from the Sperchiós River has shallowed a landlocked embayment and led to rapid progradation of deltaic land by up to 12 kilometres since 480 BC when the Pass of Thermopylae was only a narrow strip of land in front of high cliffs: it was here that the Spartan King Leonidas with a small Greek army held back a much larger Persian force in its westward advance along the coast (Fig. 46). On the coast of Elis (Western Peloponnese), where the Peneus River opens on to a shore exposed to stronger wave action from the Ionian Sea, deltaic growth has been impeded, and the sediment has been carried northward alongshore to prograde the sandy coast near Cape Glossa by about a kilometre since 1850. Similar progradation near Pirgos has resulted from northward drifting of fluvially supplied sand in the Bay of Kiparissia. In western Greece the deltaic coastal plain built by the Acheloos and Evinas

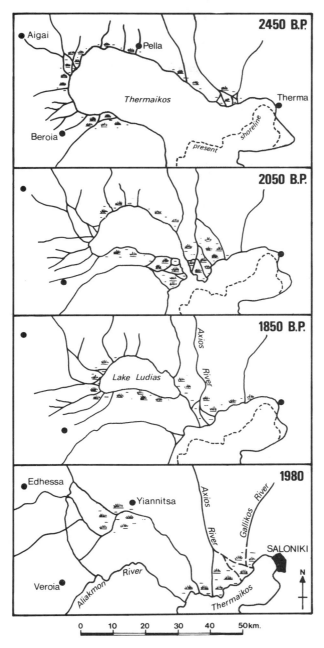

Fig. 45 Stages in the infilling of the Thermaïkós embayment, Greece, during the past 2450 years. The growth
and confluence of river deltas has stranded former ports
(such as Pella) far inland (after Kraft *et al,,* 1977)

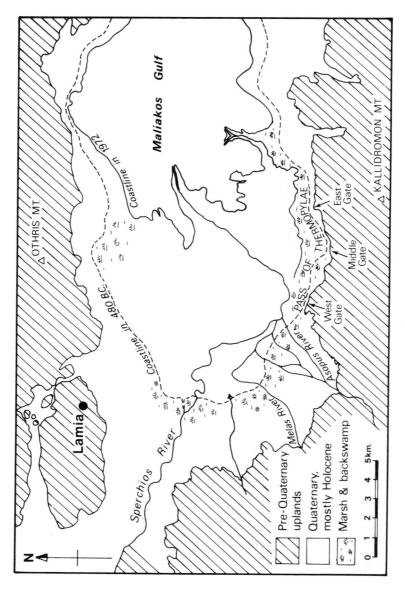

Fig. 46 Historical progradation in Maliakós Gulf, Greece, resulting mainly from fluvial deposition (after Tziavos, 1977)

Rivers has grown only slightly since it was mapped in 1838, except for a local advance of up to 2 kilometres at the mouth of the Acheloos (Piper and Panagos, 1981).

Coastline changes in Greece have been influenced by variations in water and sediment yield from rivers. As in Italy, these have been related partly to a relatively dry climatic phase during the past century, partly to dam construction and irrigation schemes, and partly to soil conservation works, including reafforestation. In addition, earthquakes have raised or lowered parts of the coastline. Kelletat (1972) has published a geomorphological map of the Peloponnese coast which shows the extent of receding cliffs and the distribution of sandy deltaic and steeper waste fan shores, many of which are prograding. Cliff recession has been rapid in silts and sandstones in the south-west near Methoni, and on the peninsulas bordering the Lakonikós Gulf.

Santorini consists of the high islands of Thera, Therasia and Aspronisi, which stand on the margins of a caldera, the site of an earlier large volcanic island blown apart by an explosive eruption about 3500 years ago: it was possibly a tsunami, generated by this explosion that destroyed Minoan settlements here and on Crete. High cliffs form the caldera walls, and recession continues on low cliffs cut in the pumice that blanketed these islands after the explosion. In 197 BC a young volcanic island (Nea Kameni) appeared in the centre of the caldera, and this has subsequently grown as the result of successive eruptions (Pichler et al., 1972). This sequence of events is similar to that seen at Krakatau in Sunda Strait, Indonesia (page 127).

Much of the island of Crete has steep coasts, in places cliffed but generally receding only slowly. Most beaches have been fluvially nourished, and are of limited extent, with some erosion reported locally near Iráklion. Tectonic movements in the late Holocene have raised some parts and depressed others (Pirazzoli et al., 1982): some ancient ports have been stranded inland as the result of uplift and deposition; others are now submerged on the sea floor.

WESTERN BLACK SEA

The sandy beaches of the south-west Black Sea, in European Turkey and southern Bulgaria, have been fluvially nourished and many of them have been eroding in recent decades. The curving beaches alongside the Pomorie tombolo are eroding rapidly. Rojdestvensky (1972) described accretion on the shores of the Gulf of Varna, where the sand has come partly from rivers and partly from cliff erosion. Much of the Bulgarian coastline is cliffed with recurrent landslides forming temporary lobes of progradation (Shuijskii and Simeonova, 1982). Locally, such lobes have acted as natural breakwaters, trapping southward-drifting sand to produce temporary beach accretion. On clay outcrops cliff recession had attained 3.5 metres per year, while the landslide lobes that formed at Saraforo, Ravda, and Kranero have been cut back more than 20 metres per year. Extensive erosion has resulted in wide-

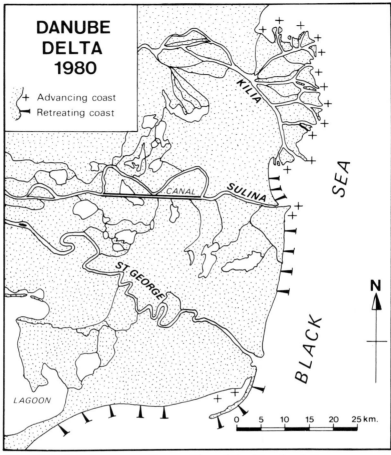

Fig. 47 Active changes on the coastline of the Danube delta, as mapped
by Gastescu and Breier (1980)

spread coastal protection works, but these were severely damaged in the
1976 and 1977 storms (Simeonova 1985).

The southern part of Romania also has receding cliffs, but erosion rates
are slower than in Bulgaria; typically about 4 metres per year between
Constantza and Agigea (Moldovanu and Selariu, 1971). Sandy beaches are
extensive, forming spits and barrier islands that have prograded, mainly from
shoreward drifting of sea floor sand, derived ultimately from the Danube.
Historical evolution of the Danube delta, to the north, has been documented
by Gaştescu and Breier (1980), who mapped stages in the growth of the Kilia
lobe between 1830 and 1979. Between 1880 and 1972 accretion alongside the
Sulina breakwater advanced the coastline up to 2.2 kilometres but to north
and south there was erosion. Active changes on the Danube delta are shown
in Fig. 47.

USSR BLACK SEA COAST

From the Romanian border to the Sea of Azov is a coastline 1628 kilometres long, of which 30% is cliffed and receding, 34% stable and 36% depositional; of the latter, some formerly prograded sectors have in recent decades become subject to erosion (Shuisky, 1985b). V. P. Zenkovich reported that sectors of active progradation were very limited on the coasts of the Soviet Union and occurred only 'in local bays, in front of some deltas, and at the terminations of large spits; in all cases the progradation is the result of huge abrasion of adjacent shores or the solid runoff from large rivers'. He noted that there are extensive areas where progradation took place in Holocene times 'just after the last marine transgression', but that many of these depositional coastlines are now either stable or retrograding.

Between the Danube delta and Odessa there is a drift divergence on either side of an actively cliffed sector at Budaki (Zenkovich, 1967), but the barriers enclosing coastal lagoons to the north (at the mouth of the Dniester) and to the south are no longer prograding, although prolongation continues at their south-western end. The major rivers have carried sediment to the Black Sea in the past, but it appears that sediment yield has diminished as a consequence of dam construction and reservoir impoundment (in the Dnieper and Dniester for example), and that coastline erosion has ensued near river mouths. The beaches near Odessa have been artificially renourished, using rocky debris brought from the hinterland. To the east, the barrier island coastline is being cut back except for the end of the Tendra Spit, which is growing at about a metre a year.

Zenkovich (1967) observed some rates of cliff recession. On outcrops of Pontic Limestone on the Odessa coast the recession had been slow (less than 3 centimetres annually), but in the southern Crimea there had been relatively rapid retreat on coasts subject to recurrent landslides (up to 9 metres per year); the landslides produced temporary capes and salients of slipped material on the shore, which were gradually consumed by subsequent marine erosion. Rapid slumping and recession have taken place on the clay cliffs of the east coast of the Sea of Azov, especially on and south of the Eysk Peninsula.

Growth has continued on parts of the spits of shelly material around the Sea of Azov (Fig. 48). In some cases there has been re-shaping of a spit, erosion of one portion being balanced by accretion on another, usually at the distal end; in others the additions result from continued arrival of shelly debris derived from marine organisms, washed in from the sea floor. Migration of projecting 'spits of Azov type' (Zenkovich, 1967) results in alternating coastline advance and retreat as the depositional structure moves past. Otherwise erosion has become predominant on beach-fringed coastlines in recent decades. Biogenic production of shelly material is of the order of 1.8 million cubic metres per year, but the supply of fluvial sediment from the Don, the Kuban, and other smaller rivers has diminished, and the proportion of sand

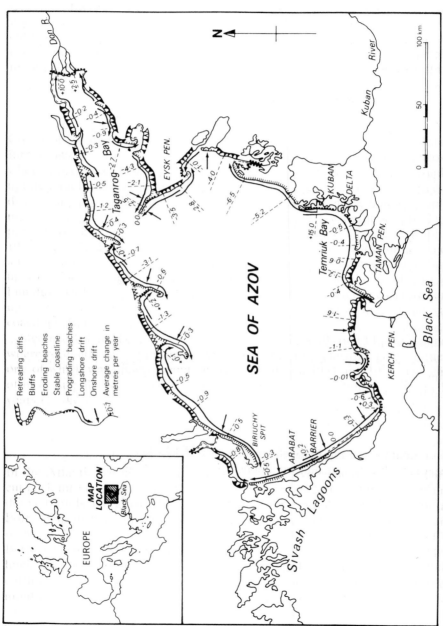

Fig. 48 Coastline changes around the Sea of Azov (based on data supplied by Y. Shuisky)

derived from cliff erosion or from sea floor rock outcrops is small (Mamykina, 1978). The coastline of the Don delta is still advancing at up to 10 metres per year, but the Kuban delta is being cut back. Cliffs are not extensive around the Sea of Azov, but those cut in clays are retreating at up to 6 metres per year, and those in limestone about 0.4 metres per year (Shuisky, 1985b).

The north-eastern coast of the Black Sea is steep and in some places mountainous on the flanks of the Caucasus Ranges. Some parts are cliffed, and occasional landslides occur on coastal slopes south-east from Anapa. Rivers deliver sand and gravel to the coast, supplying beaches which drift south-eastwards. Sediment from the Tuapse River formerly nourished beaches to Sochi and beyond, but dam construction has diminished this supply.

The building of the harbour breakwater at Sochi led to interception and local accretion updrift, and depletion of beaches downdrift, accelerated by the unwise removal of gravel from the shore for use in construction work. Severe erosion ensued, and it became necessary to replace the beach at Sochi artificially, and retain it by groynes and undersea dams (Zenkovich, 1973). There has been local progradation, continuing during the past century, in places where the longshore drift of beach material (derived from river sediment and cliff erosion) has accumulated in depositional forelands, as at Cape Pitsunda (growing at 0.1 metre per year) and Cape Sukhumi. There has been erosion on the eastern side of Cape Pitsunda because losses of longshore drift into the head of a submarine canyon has depleted the sediment supply. Erosion threatened to undermine hotels, and concrete blocks and tetrapods were dumped to halt it in 1970. Man-induced accretion has proceeded alongside breakwaters at Ochamchire (with accelerated erosion downdrift), and in the lee of obstacles such as the ships which ran aground in Sukhumi Bay in 1942 and resulted in the local building of a lobate shingle foreland which is now artificially stabilized. Cape Kodori is a large deltaic foreland augmented by sand and gravel from the Kodori River, and a smaller protrusion occurs at the mouth of the Inghuri River: between the two is an eroding sector, with cliffs cut into the deposits of an ancient delta.

At the eastern end of the Black Sea the diversion of the mouth of the River Rioni in 1939 and the construction of dams on this river have been followed by erosion of the adjacent, previously prograding, sandy deltaic coastline, which has been cut back about 900 metres, the retreat attaining a rate of up to 40 metres per year (Zenkovich, 1985). The Batumi shore is partly stable and partly retreating, some erosion being a sequel to diminished discharge from the Choronkny River.

CASPIAN COAST

The coastlines of the Caspian Sea (Leontiev et al., 1977) show the effects of emergence due to the lowering of sea level by 2.67 metres between 1930 and

1977 (Fig. 49). In addition to the gain of land by emergence of the nearshore sea floor (a gain of more than 50,000 square kilometres between 1930 and 1977) there have been sectors, notably in the south-east, where waves have washed up sand (mainly shelly and oolitic material) from the sea floor to prograde beaches and build new islands. Despite the sea level fall, some sectors are still actively cliffed (Zenkovich, 1967). In the south-west the Kura spit is still growing, but its eastern shore is being cut back following a reduction in sediment supply from the Kura River following dam construction. On the west coast, emergence has drained the Akzybir Lagoon and enclosed the Adji Lagoon in a former embayment. The sandy coast of North Azerbaijan has been stable, except for erosion on the Samur delta. About 150 years ago the Sulak river flowed into the bay behind Agrakhan spit, but it then cut a new outlet at the southern end of the spit, initiating a delta, which is now eroding as the result of dam construction and a further diversion of the river mouth southward in 1956. The Terek delta, also in the bay behind the Agrakhan spit, has been eroded since a new artificial outlet was cut southward about a decade ago: delta growth is proceeding at the new outlet.

In the northern Caspian, emergence has advanced the coastline several kilometres; locally up to 50 kilometres (Fig. 50). Emergence first facilitated the continued growth of the Volga delta (up to 2 kilometres per year between 1929 and 1945), but with dam construction and diminished sediment yield this growth has now almost ceased, and parts of the deltaic coastline are being cut back by wave action. Development of offshore sand and shell banks was aided by emergence (Zenkovich, 1967), and there were some small-scale examples of barrier initiation. Emergence has halted cliffing on the Tyub-Karagan Peninsula, leaving the old cliff stranded behind wide shelly beaches. Outlying sandy islands have become enlarged, and some have coalesced.

On the east coast a dam has been built to seal off the bay of Kara-Bogaz-Gol, thus forming an artificial coastline. To the south there has been widening and extension of spits at Krasnovodsk Bay and on the Cheleken Peninsula, as well as emergence and growth of an outlying shelly barrier island at Ogurchinskiy. Progradation has been extensive on the sandy coast of South Turkmenia, where in the absence of dune-retaining vegetation, some of the sand has been blown inland from the beach to form desert sandhills.

On the Iranian shore of the Caspian Sea emergence produced prograding beaches of dark sand, derived largely from sediment supplied by rivers draining the Elburz Mountains (Fig. 51). Predominant eastward drifting has led to the widening and growth of the spit which now almost encloses Gorgan Bay near Bandar-e-Shah, while to the west there has been a continued advance of the shores bordering the Safid Rud River north of Rasht.

After 1977 the Caspian Sea began to rise again, with the result that erosion has increased, especially on the western coastline, and on beaches and islands built up during the emergence (Leontiev, 1985). This is already a more

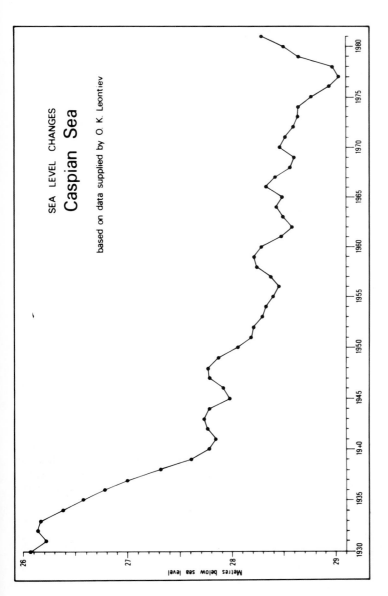

Fig. 49 The Caspian Sea fell in level between 1930 and 1977, partly because of dam construction and reservoir impoundments in the tributary rivers, and partly because of a trend toward more arid conditions in this region. Since 1977 there has been some recovery, possibly related to the damming of the mouth of the Kara Bogaz-Gol embayment on the east coast, halting the flow of the Caspian Sea into a high-evaporation area. Based on data supplied by O. K. Leontiev

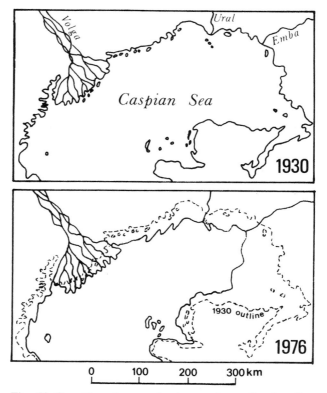

Fig. 50 Coastline changes in the northern Caspian Sea between 1930 and 1976, a period when sea level fell about 2.67 metres (from data supplied by O. K. Leontiev)

Fig. 51 Prograded sandy coastline of the Caspian Sea, Iran, as the result of emergence due to the fall of sea level that took place between 1970 and 1977 (cf. Fig 49). Photo: Eric Bird (May 1970)

substantial rise than any of the preceding oscillations (Fig. 50), and if it persists, coastal erosion is likely to become more extensive around the Caspian Sea.

TURKEY AND CYPRUS

The Black Sea coast of Turkey is generally steep, with some receding cliffs (e.g. north of Eregli), and deltas growing as the result of floodwater sediment supply to river mouths, notably the Kizilirmak and Yesilirmak Rivers. Comparison of maps made in 1916, 1955 and 1971 shows rapid progradation, especially on their eastern flanks (Erol, 1983). Sediment from the Filyos, Melen and Sakarya Rivers has drifted eastward to maintain beaches, but the supply is diminishing, and there is now some beach erosion. On the west coast, rivers draining into steep-sided gulfs have long been prograding their deltas: the Büyük Menderes, for instance, has grown into the Gulf of Miletus, cutting off the ancient port of Heracleia (Fig. 52), now partly submerged on the shores of Bafa Lake, while the Kücük Menderes has advanced its coastline 24 kilometres since the town of Ephesus was a coastal port (Kraft *et al.*, 1977). The rate of progradation has varied, however, and it is possible that increased sediment supply followed early historic deforestation and soil

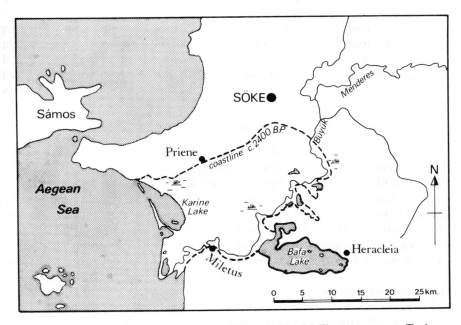

Fig. 52 The position of the coastline of the Gulf of Miletus, western Turkey, about 2400 years ago is indicated by the pecked line. Heracleia was then a port, and Priene a coastal city. Both have been stranded inland by the subsequent growth of the Büyük Menderes delta (after Kraft *et al.*, 1977)

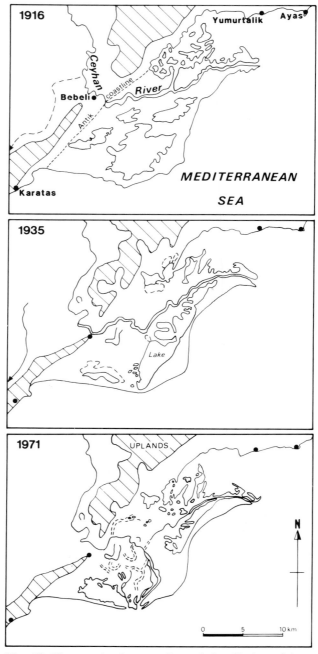

Fig. 53 Changes in the coastline of the Ceyhan delta, southern Turkey, following diversion of the river mouth southward in 1935 (after Erol, 1983). The abandoned eastern arm is being eroded, while deposition has filled the lake on the southern side, which is prograding

erosion from slopes in the hinterland has now given place to declining sediment yields from increasingly rocky catchments (Eisma, 1962).

The south coast of Turkey is dominated by high mountains and steep or cliffed coasts, but intermittently prograded deltaic plains occur at Antalya, at Erdemli-Mersin, Seyhan-Ceyhan, and in the Gulf of Iskenderun. Between 1916 and 1971 the Göksu delta, on the southern coastline, showed changes related to southward migration of the river mouth as a focus of accretion (Erol, 1983). Again, the rate of progradation has diminished in recent decades. Beach ridge construction and barrier widening are still in progress near the mouth of the Seyhan, but beach ridges built near a former outlet from this river are being cut away. On the Ceyhan delta an eastern lobe that grew until 1935 is being dissected and removed following the diversion of the river to a southward outlet, around which progradation is proceeding (Fig. 53).

Turkey is subject to frequent and strong seismic activity (Evans, 1971): in the 1958 earthquake the Black Sea coast near Bartin rose about 40 centimetres, and there is evidence of similar tectonic uplift in south-west Turkey and around the Orontes delta in the south-east (Erol, 1984).

On the island of Cyprus cliffs cut in chalk and limestone are gradually receding, especially on the north coast, but the steep coastal slopes in volcanic rock in the north-west are relatively stable. Beaches are extensive between Famagusta and the Karpas Peninsula, and intermittent along the north coast. In general they have been fluvially nourished, and some erosion in recent decades is indicated by the exposure of beach rock.

SYRIA AND LEBANON

Two-thirds of the coastline of Syria and Lebanon is steep and rocky, with cliffs cut mainly in limestones, shales, dune calcarenites, and volcanic formations. Cliff recession is slow on the ophiolite outcrops of Ras el Bassit, but more rapid on marls near Batrun. Sandy beaches back the Bay of Akkar and much of the Lebanese coast south of Beirut, and there are smaller beaches, often pebbly, at the mouths of wadis. Erosion has prevailed on these beaches in recent decades, as indicated by extensive foreshore exposures of beach rock (Sanlaville, 1984). In the Bay of Akkar, erosion followed the quarrying of sand and pebbles from the beach.

ISRAEL

Except in the steep Haifa sector around Mount Carmel (composed of Cretaceous limestone with elevations of 200–300 metres), the Israeli coast is generally low-lying. Steep sections with cliffs about 10 to 20 metres in elevation make up 23% of the coast, discontinuous quartz eolianites 22%, and wide beaches with dunes 55%. The beaches north as far as Haifa Bay contain quartzitic sand derived from the Nile (Emery and Neev, 1960; Goldsmith and Golik, 1980), with only minor contributions from local cliff erosion and seasonal streams, most of which are now dammed. North of Haifa Bay, the beach sediment is dominantly biogenic, with minor contributions from local erosion of the small, discontinuous quartz aeolianite ridges in this sector. Quarrying of sand (prior to 1963) has contributed to the narrowness of some beaches, but in general, beach and cliff erosion have been relatively minor. The cliffs of Netanya (20 to 40 metres high) have eroded approximately 40 centimetres per year from 1945 to 1978 (Ron, 1982), and studies of the 10m high cliffs south of Gaza by Goldsmith (1983) suggest recession rates of about 20cm per year. There is large spatial and temporal variability. Although offshore breakwaters along this coast have rapidly accumulated sediment, forming tombolos, the total sediment accumulation calculated by Nir (1985) was relatively minor in comparison with the annual net longshore transport. Updrift erosion and downdrift accretion have occurred alongside breakwaters at Gaza and at the small Ashqelon harbour on the north and south sides, respectively, and not at Ashdod. Although the Aswan High Dam has resulted in severe erosion on the Egyptian coast, there is no apparent effect yet along the Sinai and Israeli coasts, where the beaches continue to receive sediment eroded from the Nile delta region.

EGYPT

The Mediterranean coast of Egypt is predominantly sandy, notably along the shores of the Nile Delta, where in recent decades the long-continued progradation has given place to rapid erosion (Fig. 54). This has been most extensive on the deltaic promontories at Rashid and Damietta, which lost 1.7 and 1.8 square kilometres respectively in the interval between 1914 and 1970; the Damietta coastline has retreated by up to 40 metres per year, and the adjacent coast of Ras El-Bar has shown related erosion totalling 1800 metres between 1902 and 1960. Nielsen (1973) described erosion on the Rosetta shore, where a lighthouse 950 metres inland in 1898 was lying offshore by 1942: between 1898 and 1954 the coastline retreated 1645 metres here. Erosion has also been rapid on the central sector, on the barrier spit that stands in front of Lake Burullos (Khafagy and Manohar, 1979). The onset of severe erosion evidently began within the past century, and is probably due to the reduction of fluvial discharge and interception of sediment yield from the Nile by barrages and dams, particularly since the completion of the Aswan High Dam in 1964 (Orlova and Zenkovich, 1974). Since 1964 the Nile flood waters and their sediments have been impounded in the Lake Nasser reservoir behind the dam; only a small quantity of water is now discharged through the Rosetta estuary, and the Damietta branch is generally almost dry (Kassas, 1972).

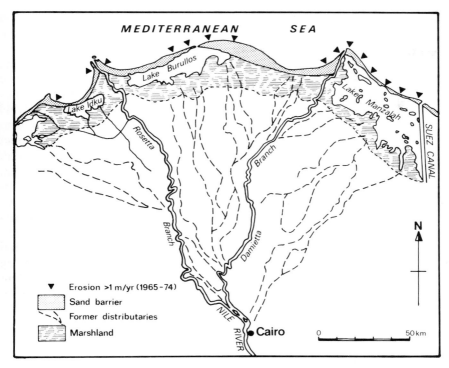

Fig. 54 Coastline recession on the Nile Delta has been extensive as a sequel to barrage and dam construction on the River Nile during the past century, and especially since the completion of the Aswan High Dam in 1964. Discharge of water and sediment to the sea from the Rosetta and Damietta distributaries has declined, and this diminished sediment yield has been followed by coastal erosion, especially near the mouths of these distributaries

LIBYA

Much of the Libyan coastline is low and sandy, with extensive sebkhas behind the shores of the Gulf of Sirte. Beach erosion has been reported near Tripoli and at Benghazi, downdrift from harbour breakwaters that have induced local accretion. There are minor receding cliffs near Tobruk and Ras Hilal, and to the east of Tripoli (Schwartz, 1985b).

TUNISIA

Beaches are extensive along the Tunisian coastline, with only limited sectors of low, receding cliff, and some higher cliffs on the north coast west from Bizerta. Much of the coastline is receding, and some archaeological sites have been lost. Of 484 kilometres of sandy coastline, 10 kilometres are advancing, 360 kilometres stable and 115 kilometres retreating. Beach erosion, often initiated by sand quarrying, is very widespread (Paskoff, 1981b), and severe on Djerba Island, where some of the eroded material has drifted north-west to extend the spit at Ras Rmel by 250 metres between 1963 and 1972, and more than 2 kilometres since 1885 (Miossec and Paskoff, 1979). There has also been erosion of sandy beaches formerly nourished by shoreward drifting from the sea floor in the Gulf of Gabès and the Gulf of Hammamet, and notably in the Gulf of Tunis, where some of the sand came from the Miliane and Medjerda Rivers. At Bizerta, sand drifting from the north has accumulated alongside the harbour breakwater, advancing the coastline up to 250 metres (Mathlouti and Paskoff, 1981), and there has been similar progradation alongside the northern breakwater at Ghar el Melh. The growth pattern of the Medjerda delta was modified following its cutting of a new outlet during a major flood in March 1973. Deprived of a sediment supply, the old delta has been eroded, derived sandy material accumulating in the southward-growing spit of Foum el Oued, while a modern delta is developing at the new outlet, in its lee (Paskoff, 1978, 1981c). Longer-term changes have stranded the port of Utique, which was still active in the fourth century AD, 10 kilometres inland behind the Medjerda delta, whereas the foundations of the ancient port of Carthage are on the sea floor in front of an eroding beach at Tunis, at a site where sea level has risen about 2 metres since the Roman era. On the east coast of the Cape Bon Peninsula, rivers dissecting soft Tertiary sandstone catchments are still delivering sand to maintain the adjacent beaches.

ALGERIA

Beaches alternate with cliffed and rocky shores along the Algerian coastline, occupying embayments or coves, particularly at the mouths of wadis. Minor accretion has occurred where sand is washed down from wadis to the shore, as at Sidi Youcef, and between 1882 and 1973 sandy forelands prograded up to 250 metres on either side of the Oued Isser at Kabylia. Sand drifting eastward has also accreted alongside harbour breakwaters, as at Zemmouri, east of Algiers, and inside new harbours as at Sidi Fredj, west of Algiers. Cliffs cut in various formations, especially Pleistocene dune calcarenites, have been retreating, as at Tipaza, west of Algiers (Mahrour and Dagorne, 1985).

MOROCCO

The Mediterranean coastline of Morocco consists partly of slowly retreating cliffs in dune calcarenite and Rif formations, and partly of sand and pebble beaches, fed mainly by wadis. There has been erosion on the sandy coasts, especially east of Nador. On the Atlantic coast of Morocco sandy beaches are more extensive, especially along the Plains of the Rharb and the Sous, the intervening sector having cliffs and bluffs, as at Cape Cantin. NNE trade winds sweep desert sands to the shore, nourishing beaches at Cape Sim, near Essaouira, and locally along the Saharan coastline to the south (Weisrock, 1985).

WEST AFRICA

Progradation of sandy coast south from Rio del Oro, in Saraoui, has resulted from Saharan dune migration to the shore, but the cliffed coasts north to Cape Bojador and south to Cape Blanc are retreating. The sandy beaches which fringe Saraoui and Mauritania (where Saharan barchans reach the coast between Levrier Bay and Cape Tafarit) are subject to southward drifting, which has built the barrier spit (Langue de Barbarie) that deflects the mouth of the Senegal River southwards. The changes that have taken place here within the past century include longshore growth of the spit, but there is no evidence of progradation on the sandy beaches of the ocean coastline here (Guilcher and Nicolas, 1954), or south to Cape Verde.

South of Cape Verde the long estuarine inlets of the Gambia and Casamance Rivers are mangrove-fringed, and from here southwards mangroves have locally advanced on mudflats in sheltered inlets and embayments on a low wave energy coast. Intermittent sandy shoals and cheniers mark stages in progradation, as at Pointe aux Oiseaux, where a spit is growing southwards. Similar features are seen along the generally prograding mangrove-fringed coastlines of Guinea and Guinea Bissau, especially on the growing islands off the Geba estuary.

In Sierra Leone there are retreating cliffs on soft rock outcrops south of Freetown and at Shenge, and the sandy coastline begins at Sherbro Island, with progradation continuing on Turners Peninsula, which has multiple beach ridges. Sandy beaches fringe the Liberian coastline, with intermittent rocky sectors. Erosion has been reported on the seaward margins of coastal barriers between Cape Mount and Monrovia, near Buchanan, and at Greenville, where there has been minor beach accretion alongside a breakwater.

On the Ivory Coast, the western portion has remained generally stable, the central sector between Sassandra and Grand Bassam has a retreating sandy coastline (except where the breakwater at Vridi has intercepted eastward drift to produce local accretion), and the barrier coast east of Grand Bassam has prograded in recent years. Le Bourdiec (1958) concluded that these sandy barriers had received sediment from the floor of the Gulf of Guinea, but present changes can be explained by eastward drifting of coastal sands with accretion west of the harbour breakwaters at Abidjan (Hinschberger, 1985). In 1903 the beach to the east of these breakwaters suddenly retreated up to 80 metres in 35 minutes following nearshore submarine landslides into the head of the nearby Trou sans Fond.

There is a similar pattern in Ghana, where comparison of coastal surveys of 1945 and 1973 shows that progradation continued only in sectors adjacent to breakwaters (New Takoradi, Nyiasia) and rocky headlands (Apam) where the eastward drift of beach sand had been intercepted. Minor accretion was also detected on the ends of spits bordering lagoon entrances and river mouths, notably at the Volta mouth near Ada, but the Volta delta coastline has been retreating for several decades. Progradation in this period was confined to about 2 kilometres of the 600 kilometre Ghanaian coast. Sandy coastlines that had previously prograded are now in retreat, with beach rock exposed in places up to 45 metres offshore (Dei, 1972). Retreat is also in progress on cliffs cut in soft sedimentary rocks and deeply weathered igneous rock outcrops.

Eastward drift of sand continues along the generally eroding coastlines of Togo and Dahomey, and has been intercepted by the breakwater built in 1900 to protect the entrance to Lagos Harbour in Nigeria, so that Lighthouse Beach, to the west, has prograded up to 30 metres per year and Victoria Beach, to the east, retreated by a similar amount (Usoro, 1985; Webb, 1960) (Fig. 55). Pugh (1954) noted that earlier progradation, marked by beach ridge construction, had recently given place to retrogradation of Nigerian sandy shores. Local progradation has continued, however, on parts of the large lobate delta of the Niger, especially in the sector of longshore drift convergence near Mahin to the west.

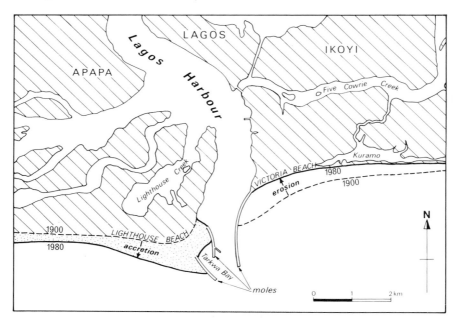

Fig. 55 Since breakwaters were built in 1900 to stabilize the entrance to Lagos Lagoon, longshore drifting of sand from west to east has resulted in progradation on Lighthouse Beach, to the west, and erosion on Victoria Beach, to the coast
(after Usoro 1985)

The Cameroons coastline has prograding mangrove swamps interspersed with sandy beaches, except for the receding cliffs between Idenao and Victoria, cut into the volcanic rocks of Mount Cameroon. To the south there has been slow progradation on the coastline near the Sanaga River, where sediment drifts north to prolong the spit at Souelaba Point, and at the mouth of the Nyong. Beach erosion has been prevalent along the southern coast, past Kribi to Bata in Equatorial Guinea, and there are receding cliffs on Cabo San Juan. Removal of beach sand for building construction has contributed to shore erosion at Libreville, but accretion has continued on the longshore spits north of the Congo as far as Cape Lopez, where sand is drifting northward. There are minor cliffs and rocky outcrops at Mayumba and Pointe Banda in Gabon, at Pointe Indienne and Pointe Noire in Congo, at Pointe de Tafe in Cabinda, and Moando, just north of the Congo River in Zaire (Diresse and Kouyoumontzakis, 1985).

On the Angolan coast, accretion has continued on the northward-growing spits off the mouth of the Congo, and at Pointado, Mussulo, Lobito, Porto Alexandre and Ponta de Marca, off Tiger Bay (Fig. 56)—the latter was breached at its southern end in 1962, and is now an island (Guilcher et al., 1974). On the inner shore of Tiger Bay are numerous cuspate spits, fed with sand from desert barchans arriving at the coast. Lobito spit has received sand from the Catumbela River, but the sediment yield has diminished because of dam construction, and there is incipient erosion. The Palmeirinhas spit is still growing, but the spit at Luanda, to the north, is withering as the result of a diminishing sand supply: it is now in the lee of the Palmeirinhas spit (Guilcher et al. 1974). Intervening sectors of the Angolan coast show receding cliffs cut in a variety of sedimentary formations.

In Namibia (South-West Africa), Bremner (1985) described a number of sectors of sandy coastline progradation. South of Conception Bay, a cuspate foreland has prograded in such a way that a ship (the *Eduard Bohlen*) wrecked on the beach in 1912 lay 500 metres inland by 1978. There has also been growth of the spits in Walvis Bay (where Pelican Point has been prolonged about 1.8 kilometres in the last 80 years), at Sandwich Harbour, Conception Bay, and Meob Bay, each related to the northward drifting of coastal sands. Present-day sources of sand are partly fluvial (light brown sands from the Orange River) and partly aeolian (where southerly winds drive red Namib Desert dunes into the sea, as to the south of Lüderitz), but Bremner has suggested that these four spits may be related to the reworking and shoreward drifting of sand from the submerged Pleistocene deltas of rivers that now fade into the Namib Desert: the Swakop, the Kuiseb, the Tsondab and the Sossus. The Swakop, which still flows intermittently to the sea, built a delta during the 1934 flood, but this has since been removed by erosion. Intervening sectors of cliffed and rocky coastline have been eroded, and unvegetated coastal sand formations are very extensive. Local and temporary progradation has occurred where backshore sands have been excavated in search of diamonds on the Namibian coast north of the Orange

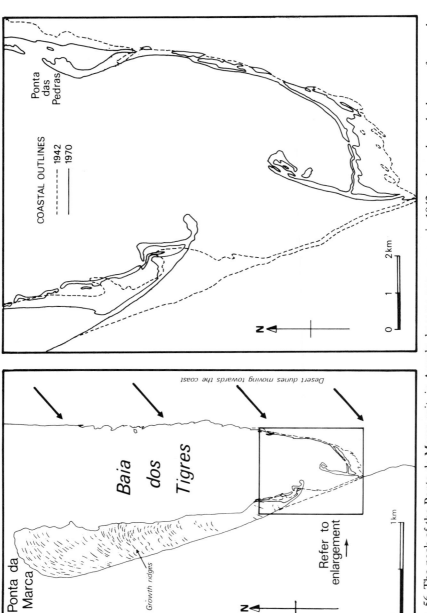

Fig. 56 The neck of the Ponta da Marca spit, in Angola, became very narrow in 1942 and was breached soon afterwards. By 1970, paired spits had developed on either side of a wide gap leading into the Baia dos Tigres (Guilcher, 1974)

River. Overburden is bulldozed into the sea to form an embankment which is gradually cut away by wave action until the original profile and alignment have been restored.

SOUTH AFRICA

South of the Orange River the coast becomes rocky and irregular, and there are receding cliffs. Beaches become more extensive between Olifants and Britannia Point, the general trend being erosional, although sand blown from the hinterland is being deposited to prograde the beach locally at Spoegrivier and at the Groen estuary. There has been little change on the hard granitic shores to the south, but at Saldanha the harbour works include an artificial beach emplaced to prevent erosion. Heydorn and Tinley (1980) found that accretion was in progress on the beaches of St Helena Bay, north of Bergrivier, and noted that this was unusual in terms of the modern prevalence of erosion on South African beaches.

Changes have been slow around the sandstone coasts of the Cape of Good Hope and Cape Hangklip, but the sandy coastline at the head of False Bay has been cut back in recent decades. Minor beach erosion has occurred in the sandy bays along the Hermanus coast and near Cape Agulhas. The larger Cape Coast beaches, occupying asymmetrical embayments between rocky headlands, are backed by extensive dunes. Sand driven eastwards by wind action has spilled across the headlands into adjacent bays, and there has also been shoreward movement of marine sand from the sea floor on to beaches. Local accretion has occurred around lagoon entrances and river mouths, as at Stilbaai, where the coastline has receded between 1855 and 1964, but has since prograded since the arrival of sand from the sea floor. At Plettenberg Bay a lagoon outlet formed by a 1915 flood has since migrated southwards along the enclosing barrier, but the coastline has otherwise changed very little. At Gamtoos sand washed in from the sea floor has closed an abandoned river outlet. Wind-blown sand spilling over Cape St Francis formerly maintained beaches in the bay to the east, but after the planting of dune vegetation diminished this supply, beach erosion ensued. The same sequence has occurred on Cape Recife, with beach depletion east of Port Elizabeth. Sand lost from these beaches has drifted alongshore to accrete against the southeastern breakwater at Port Elizabeth, an interruption which has, in turn, led to erosion of beaches to the north-west countered by the emplacement of about 6 kilometres of tetrapods (Fig. 57). Mobile dunes advancing eastwards along the coast are spilling into the deflected estuary Sundays River, and at Port Alfred drifting beach and dune sand has prograded west of the harbour entrance breakwaters, with erosion developing on their eastern side (Heydorn and Tinely, 1980). The beach and dune fringe extends intermittently north-eastwards past East London to Durban, without any naturally prograding sectors. At Durban, northward-drifting sand has accumulated

Fig. 57 The beach that formerly fringed the coastline north of Port Elizabeth, South Africa, has been depleted as a sequel to harbour breakwater construction, which has cut off the northward-drifting sand supply. Erosion ensued, and the coastline has been stabilized by dumping concrete tetrapods. Photo: Eric Bird (January 1983)

alongside the harbour breakwater, and the beach in Durban Bay to the north has been depleted.

However, near Mtunzini there is a 2 kilometre sector that has prograded about 95 metres between 1937 and 1977, described as 'one of the few coastlines in Natal where sand deposition and foredune advance is occurring' (Weisser *et al.*, 1982): the sand has evidently come from adjacent rivers. The coastline at Richards Bay harbour has been modified by accretion alongside breakwaters and by land reclamation at the entrance to Mhlatuzi Lagoon, while the St Lucia lagoon entrance changed frequently between 1905 and 1972, and was then stabilized by breakwaters: subsequently there has been accretion on their southern side and erosion to the north (Orme, 1973).

MOÇAMBIQUE

The beaches of northern Natal continue to Santa Maria Cape in Mocambique, which partly shelters the Bight of Maputo, where swampy shores are prograding. Some deltaic accretion has occurred around the mouth of the Limpopo River, but in general the sandy coastlines have retreated along the Inhambane sector, and in the Bight of Sofala there are fringes of dead mangroves, undercut shores, slumped dune vegetation and cliffed beaches (Tinley, 1985). The Zambezi delta has prograded during the past century, but there has been some erosion of its coastline since the completion of the Kariba Dam in 1958. Beach ridges have been added to the coast south of the Luria River mouth, and there has been minor progradation at the mouth of Rovuma River.

MADAGASCAR

In Madagascar there has been little change on the east coast, where hard rock formations and low sandy coastlines predominate in the central parts, and sandy coastlines in the south-east and north-east. The main changes have occurred on the sandy and swampy sectors of the west coast where accretion has continued around and between the mouths of the Mangoky, the Morondava, the Tsiribihina, the Manambolo and the several rivers of the north-west. There has been growth of sand spits, for instance at Sarodrano, on the south-west coast near Tuléar (Battistini and Le Bourdiec, 1985). On the other hand, erosion of the coast at Morondava has necessitated groyne construction to retain the beach.

TANZANIA

Alexander's (1966) descriptive classification of the north-east coast of Tanzania provides basic data that are otherwise rare on the African coastline, and his study of beach ridges in the same area (Alexander, 1969) indicated a history of alternating erosion and deposition. The sand has come mainly from the rivers, and has drifted northwards along the shore. Progradation continues on deltas, but elsewhere erosion is predominant, destroying coastal coconut plantations: it is thought to have started between 40 and 140 years ago. On the Kunduchi coast, north of Dar-es-Salaam, parallel beach ridge terrain has been cut back about a metre per year by waves arriving obliquely from the south-east, producing longshore drift north to the promontory of Ras Kiromani, and thence to an accreting shore in Unonio Bay. Slow erosion is in process on cliffs cut into emerged coral, especially around offshore islands. Off northern Tanzania the sandy island of Mawizi suffered erosion, and disappeared completely in 1982.

KENYA

Much of the Kenya coastline consists of emerged coral reefs, with slowly retreating cliffs at their seaward margins, and intervening beaches generally backed by eroding dunes or beach ridges, and fronted by shallow lagoons and fringing reefs, as at Nyali (Bird and Guilcher, 1982). At Malindi the seaside resort had beach progradation of up to 150 metres between 1975 and 1981, sand having drifted south from the mouth of the Galana River: previously it drifted mainly north to nourish beaches out to the headland of Ras Ngomeni. Further north, F. Ojany reported that sand from the Tana

River has prograded the coastline near Kipini, and there has been an advance of mangrove shores in the sheltered bays behind Lamu.

SOMALIA AND DJIBOUTI

Although there are extensive mobile sands in the coastal fringe of southern Somalia the prevailing south-easterly winds drive them inland, and the beaches are generally either stable or retreating. Occasional floods deliver sand from the Juba River to the coast near Chisimaio, but this is dispersed, mainly northwards, by waves and currents and there has been little sustained progradation. Sand washed from wadis has nourished local beaches, and produced accretion on the Ras Hafun tombolo. On the Gulf of Aden coast, westward drifting of sand has resulted in local progradation and the growth of spits at Khor Shori and Berbera. At Zeila a cuspate spit is growing out from the coast, and there have been minor gains and losses on Duren and Kolokhtiya Islands, which are breached residuals of an earlier spit close to Djibouti.

RED SEA COASTS

Much of the Red Sea coastline is low-lying or hilly and reef-fringed. Beaches are generally narrow and intermittent, except where they have been augmented by coralline sand washed in from reefs, or by deposition of flood outwash from wadis or infilling at the head of marine inlets (sharms). Locally, slow erosion is in progress on cliffs cut in emerged coral, as on the emerged Abulat Knoll on the Farasan Bank (Guilcher, 1985), but in general there have been only minor changes on the Red Sea coastlines of Ethiopia, Sudan, Egypt, Israel, Jordan, Saudi Arabia and Yemen in recent decades.

SOUTHERN ARABIA

In the Aden district, cliffs formed in volcanic formations are retreating at varying rates, those cut in volcanic ash and agglomerates more quickly than those cut in lavas. Intervening sandy beaches have been generally stable, with episodic accretion near wadi mouths after occasional floods. Longshore drifting to the west is prolonging spits, for example at Khawr am Umayrah. To the east there are receding cliffs at Ras al Mukalla, between Shihr and Sayhut, at Ras Fartak in the People's Republic of Yemen, and at Ras Madhraka and Ras al Hadd in Oman. Offshore, the island of Socotra has rugged cliffs of granite and limestone, and on the north coast sandy beaches, mainly near the mouths of wadis: changes have been limited and localized. Kuria Muria Islands are also hard and rocky, with only gradual changes in progress, and the same is true of Masirah Island, in the lee of which the Hikman Peninsula is prograding as sand is washed in from nearshore shoals.

At Ras al Hadd the limestone cliffs continue north-westwards, past Muscat, where banks of concrete tetrapods protect the coast road, to Matrah, where sandy beaches fringe the Batina coastal plain. These are generally stable, with occasional local augmentation from wadis during rare floods. Northwards, cliffs cut in limestone are retreating slowly, but on the Musandam Peninsula there is local rapid retreat and landsliding where coastal outcrops of shale beneath the limestone are being undercut. Gravelly beaches occur at wadi mouths, and some of these have intermittently prograded.

ARABIAN GULF COASTS

The south and south-west shores of the Arabian Gulf show extensive sebkhas (low-lying sandy and muddy depositional flats slightly above normal high tide level), bordered by spits, some of which are prograding. Calcareous biogenic sands are forming in the nearshore area, and producing extensive shoals (Great Pearl Bank) from which sand has moved onshore to beaches near Dubai and barrier islands near Abu Dhabi (Purser and Evans, 1973). Shinn (1973) described intertidal biogenic carbonate sedimentation on the east coast of the Qatar Peninsula, where southward drift has nourished spit growth, for example on Ras Umm Sa, which grew rapidly between 1957 and 1965. Farther south, desert dunes moving across the Umm Said sebkha are spilling into the sea and prograding the coastline by deposition of terrigenous quartzose sand. There has been little change around Bahrein Island during the past century, apart from the extensive land reclamation on the north coast at Al Manamah. On the Saudi Arabian coast progradation has enlarged the sand spit at Ras Tanurah, and there has been some accretion alongside harbour breakwaters to the north. Minor cliffing is seen on calcareous rocks at intervals along this coast and in Kuwait, with slow progradation on the sandy shores of the Gulf of Kuwait.

At the head of the Gulf the Tigris-Euphrates-Karun delta is an area of continuing sedimentation in coastal swamps, but archaeological evidence suggesting rapid progradation (Beke, 1834) has been questioned (Lees and Falcon, 1952), and it is possible that tectonic subsidence within historical times has prevented much advance (Berry et al., 1970). Mangrove progradation has been reported on the north-west of Bubiyan Island.

IRAN

The northern deltaic coastline of the Arabian Gulf continues eastwards into Iran, and includes the Zohreh delta, the southern part of which is prograding. Beaches fringe the coast south to the swampy prograding deltas of the Hilleh and Mand Rivers near Bushire, but the outlying island of Kharg has cliffed coastlines. Between Bushire and Bandar-e-Lengeh, beaches fed by rivers and wadis alternate with receding cliffs, and offshore islands and are also cliffy. The Mehrar delta is prograding in the lee of Qeshm Island, and there are beaches built forward by sediment from wadis near Bandar Abbas, on the east coast of the Strait of Hormuz, and along the Makran coast, where stronger ocean swell has shaped bay beaches between cliffed sectors (Sanlaville, 1985a). Between Jask and Bandar Abbas extensive mudflats are backed by clumps of low mangroves.

PAKISTAN

Snead (1970) showed that the Makran coast of Pakistan has been subject to tectonic uplift, which has continued into recent decades: in the 1945 earthquake the coast near Pasni was raised 4.6 metres, and two rocky shoals emerged as new islands. In the west, the Dasht River brings little sediment to the coast, and along the Makran the beaches are derived either from cliff erosion or from sediment washed in from the sea floor: most streams sink into sand before reaching the shore. The Hab River, near Karachi, delivers sand to form offshore shoals, from which wave reworking drives sediment onshore to beaches. Cliffs cut in Tertiary sandstones and Pleistocene deposits alternate with sandy beaches, many of which are receding, but it is thought that spasmodic uplift in recent centuries led to earlier phases of sandy deposition. Progradation on uplifted sectors has been followed by erosion, and wind-drifting of sand inland in the form of barchans, as on the tombolo at Ras Ormara. Erosion has predominated, despite the recurrent uplift of long sectors of this coastline, the only sustained progradation being near the entrance to Miani Lagoon, near the town of Sonmiani. Here sand deposition and the spread of mangroves on to accreting mudflats at the mouth of the Windar River during the past century have impeded boat access to Sonmiani, which was a busy port in the eighteenth and nineteenth centuries, and has since declined (Snead, 1970).

At Karachi there are retreating cliffs, and to the south-east there has been continuing progradation of salt marshes and mangrove areas adjacent to the mouths of the active distributaries in the central sector of the Indus delta, with accretion locally up to 25 metres per year. Within historical times this delta has grown to fill a previous embayment, and comparison with maps made in 1887 shows that parts of the delta coastline have advanced up to 3 kilometres since that date. In recent years there have been gains of up to 25 metres per year. The abandoned deltaic plains which flank it to the northwest and south-east have been subject to erosion, possibly accentuated by recurrent tectonic subsidence.

INDIA

According to Ahmad (1972) about 55% of the Indian coastline is beach-fringed, the beaches having been generally either stable or receding in the past few decades; about a quarter of the Indian coastline is still receiving sediment supplied from rivers. On the west coast there have been episodes of submergence due to earthquake subsidence (as in 1819 and 1845), followed by alluviation and progradation, particularly around the Gulf of Kutch. The

Gulf of Cambay is an area of continuing swampy progradation, and around and between the estuaries of the Narmada, the Tapti, and the Ulhas, farther south, there is a large sediment discharge derived partly from gully erosion in the coastal plain. Gradual accretion of silt and clay has been observed on islands at the mouth of the Narmada (Bedi and Vaidyanadhan, 1982).

On the west coast there are steep and rocky sectors, some with active cliffs, others with slopes plunging directly into the sea. In Goa, Wagle (1982) concluded that the coastlines was 'prograding along the beaches while retreating along the cliffs and headlands', but on the sandy Kerala coast there has been sustained recession, notably in the Quilon, Alleppey and Calicut districts. Beaches on the Coromandel coast typically front cliffed dune margins. The chief prograding sectors are on the deltas of the east coast, where the Krishna, the Godavari, the Mahanadi, and the Ganges have all been supplying sediment (mainly silt) to the coast. Deltaic shorelines are here generally muddy, but where sand is present beaches and spits have been formed. Spit growth on a smaller scale has accompanied the evolution of the Krishna delta between 1928 and 1978, general progradation having been interrupted by erosion of the coastline during the 1977 cyclone (Nageswara Rao and Vaidyanadhan, 1979). On the Godavari delta, northward drifting of sand has extended a sandspit at Kakinada Bay by several kilometres during the past century (Sambasiva Rao and Vaidyanadhan, 1979). Northward sand drift on this coast is also indicated by the accumulations on the southern sides of breakwaters at Madras and Visakhapatnam. On the Mahanadi delta there has been slight overall progradation by fluvial deposition, although in detail the shoreline has shown alternating advance and retreat, while to the north the coastline bordering the Brahmani River mouth has recently started to erode (Meijerink, 1982). The Balasore coast, north from Palmyras Point, is also thought to have prograded by deposition during the past century (Nagaraja, 1966).

SRI LANKA

The coasts of Sri Lanka are generally low-lying, with long sandy beaches and barriers backed by lagoons, estuarine and deltaic areas, and swamps. Steep coasts are limited mainly to the south of the island between Dondra Head and Tangalla, the Trincomalee area to the north-east and the Kudremalai Hills area on the west coast. On the south coast, the beaches are interrupted by protruding headlands of Archaean igneous and metamorphic rocks. Swan (1982) reported that beach erosion has been prevalent. Between Colombo and Dondra Head, beaches occupy 95% of the coastline; beach erosion has been spasmodic, often severe, along two-thirds of this shore, which had previously prograded. The recession which now prevails is a comparatively recent phenomenon, but no dates are available.

The main sources of beach material in south-west Sri Lanka have been from rivers, cliffed headlands and eroding coastal sands of older, higher beaches. During the vigorous and pluviose south-west monsoon, rivers in spate have disgorged large quantities of sand into the sea, and the ensuing shoreward creep of this sand during the north-east monsoon is the chief source of beach nourishment. There is evidently now an insufficient sand supply to maintain beach alignments. On some sectors coastline recession has isolated a number of headlands as offshore islands. 'Such headlands originated as residual blocks of crystalline rock which came to be woven into the fabric of the coast during an earlier phase of progradation following the Holocene rise of sea level. Since the mid-nineteenth century the soft sands, alluvia, coral or sandstone circumscribing headlands have been eroded away, thereby isolating them' (Swan, 1965). An example of this is quoted from Seenigama, about 100 kilometres south of Colombo, where the isolation of a headland in 1921 has been followed by rapid erosion of the soft sandy hinterland between Hikkaduwa and Dodanduwa; where seaside hotels have lost their beaches. Coastline recession has also been rapid between Maggona and Beruwala, to the north, where boulder walls have been put in to preserve the coastal highway (Swan, 1982).

Swan also mentioned the extraction of beach material and the quarrying of coral and sandstone reefs fringing the shore as practices which have accelerated erosion, and noted that in recent years the building of sea walls, groynes, and boulder revetments has locally stabilized the coast, and in places achieved minor progradation. On the north-west coast there has been accumulation of northward-drifting sand to prograde sectors of barrier spits near Puttalam, and on the east coast similar northward drifting has deflected lagoon outlets and prograded beaches in the north, towards Point Pedro.

BANGLADESH

Progradation has continued on parts of the deltaic coastline of the Ganges-Brahmaputra as the result of continued fluvial sedimentation, but there has also been much erosion, especially during storm surges (La Fond, 1966). Comparison of Landsat imagery in 1972 and 1979 showed intricate gains (totally 932 square kilometres) and losses (totalling 717 square kilometres) along low-lying mangrove-fringed shores and around deltaic islands (Polcyn, 1981). Much of the coastline to the east is fringed by low, narrow sandy beaches, extending around Sandwip and the other deltaic islands east of the Maghna River. Sectors of receding cliff occur west of Chittagong and along the coastline south of Cox's Bazaar (Snead, 1985).

BURMA

Mangroves are extensive in the estuarine inlets along the west coast of Burma south to Point Negrais, and many sectors have shown seaward advance in recent decades, especially where river loads have been increased as the result of soil erosion in their hilly catchments. Earthquakes have resulted in uplift or subsidence locally: Cheduba Island showed coastal emergence as a result of a 1762 earthquake. Sand spits have grown to deflect river mouths and facilitate mangrove spreading at Cypress Point and the mouth of Sandoway River. Offshore, mud volcanoes have formed temporary islands, soon planed off by wave scour, notably in Cheduba Strait.

There has been continued progradation on the central sector of the coast-line of the Irrawaddy delta south-west of Rangoon. An advance of up to 6 kilometres per century (land gain 10 square kilometres per year) has been reported from this area (Volker, 1966).

On the Tenasserim coast there has been little change on rocky promonto-ries, but intervening bays, especially at river mouths, have shown rapid silt accretion and mangrove encroachment. Quarrying of hillsides and valley floors for minerals and precious stones has increased fluvial sediment yields and accelerated this accretion. In southern Burma and the Mergui archi-pelago steep coasts predominate, and there has been little change in coastal outlines in recent decades.

WESTERN THAILAND

South from the Burmese border to Phuket the island-fringed steep coast of Thailand shows cliffing, and there has been slight recession in recent decades on west-facing sectors. Some beaches in embayments have eroded, exposing beach rock; others have been depleted by the extraction of placer tin deposits. Minor retreat has also occurred around many of the basally notched limestone tower karst islands in Phangnga Bay. The low-lying coastline between Phangnga and Satun, on the other hand, is fringed by extensive mangrove forests, in places advancing on to tidal mudflats. Sectors facing west or south-west are beach-fringed.

MALAYSIA AND SINGAPORE

The west coast of Malaysia has sectors of coastal plain with extensive mang-rove progradation continuing near the mouths of rivers such as the Kedah, the Muda, and the Perak. Tjia (1973) quoted Koopman's measurements of 100 metres of coastal accretion near the mouth of the Perak between 1881 and 1961. Steep coasts, locally cliffed, occur on Penang Island and the Langhkwai Islands, and mangroves in estuaries and sheltered embayments, as on the west of Penang Island. Sandy beaches are found only intermittently on the west coast, close to eroding outcrops of weathered granite or sandstone.

Singapore Island has some steep coast, with cliffing in weathered rock outcrops, the remainder being depositional, with beaches of sand and lateritic gravel, and areas of mangrove. The coast of Singapore Island has been substantially modified by man in the past 20 years. Land reclamation projects and port development have advanced the coast, especially in the south-west and south-east, and a substantial stretch of the new coast is protected by revetments, sea walls and breakwaters (Fig. 58). Mangrove swamps dominate whatever is left of the unmodified northern and more sheltered coast along the Straits of Johore; there are some active cliffs at Changi, Punggol and Lim Chu Kang and sandy shore stretches at Sembawang, Seletar and Changi. Until their reclamation in the 1970s the outlying islands had a high percentage (30%) of steep coast with some cliffing in weathered rock outcrops: the remainder was depositional with beaches of sand and lateritic gravel and areas of mangrove (Swan, 1971). Dumping of dredged sediment has enlarged several islands and created new ones, as at Pulau Ular and Terumbu Retan Laut. On the unmodified outlying islands, cliffing and beach and spit accretion proceed on the more exposed sides, while swamp encroachment is more common on the sheltered sides.

On the east coast of Malaysia steep sectors without beaches comprise 7.7% of coastline length, the remaining 657 kilometres having beaches, in places at the foot of receding cliffs. In addition to longshore drifting (Raj, 1982) these beaches show seasonal changes related to monsoon conditions. They are cut back during the north-east monsoon (November-February) which generates erosive waves and produces river floods and substantial sediment yields. During the south-west monsoon (May-August) beach accretion takes place through landward-migrating sand bars which move up on to the beach

Fig. 58 Parts of the coastline of Singapore Island have been prograded by artificial reclamation. The former coastline is seen in the background, and an attempt has been made to stabilize the prograded sandy coastline with the aid of intermittent breakwaters, intended to produce a series of asymmetrical 'equilibrium beaches'.
Photo: Wong Poh Poh (June 1972)

(Wong Poh Poh, 1981). Superimposed on these short-term changes are longer-term progradation of beaches and new beach ridges as at Merang, or retrogradation, where beach-ridge plains built up in earlier stages are being trimmed back (Swan, 1968). Coasts where river catchments are small, or sandy sediment yields from streams low, are subject to erosion, notably at Kuala Dungan and to the south of Sungei Jemaluang: a local exception is Jasons Bay, which receives the rivers Sungi Sedili Besar and Sedili Kecail. Nossin (1964) reported evidence from maps and charts indicating the infilling of an open embayment that existed south of Kuantan River in 1726, and traced stages in the progradation of the northern part of the Pahang delta since early in the seventeenth century (Nossin, 1965): in recent years, coastal sedimentation has increased because of the augmented yields from streams draining valley floors that have been disturbed by the dredging of alluvial tin. Deltaic coastlines at the mouths of the Endau, the Terengganu, and the Kelantan rivers have prograded by the advance of mainly swampy fringes, with some sandy shores, but there have also been episodes of recession, especially on sectors away from active sedimentation around migratory river outlets. Regrettably, the Beach of Passionate Love (Pantai Cinta Berahi) near Kota Baru is being consumed by erosion: chalets that were 180 metres from high tide line in 1971 were only 70 metres inland by 1978. Koopmans (1972) traced the changing outline of the Kelantan delta between 1943 and 1966, dominated by the westward growth and migration of the large sand spit, Pantai Laut, which has curved round to attach itself to the coastline, thereby enclosing a lagoon in the Bay of Tumpat (Fig. 59). Thus the beach at Tumpat, where the Imperial Japanese Army landed to invade Malaya in December 1941, was cut off from the South China Sea. Subsequently the

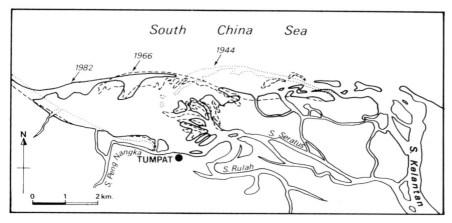

Fig. 59 Growth of the Pantai Laut sand spit, on the northern shore of the Kelantan delta, Malaysia, since 1944, has proceeded in such a way as to reach the coast to the west of Tumpat (after Koopmans, 1972, with additions from Teh Tiong Sa). The spit has been breached north of Tumpat, and the detached sand mass is now likely to drift westward along the coast

looped spit has been breached, and the detached portion is likely to be re-shaped as it migrates westward along the coast towards the Thai border.

Progradation has also occurred on parts of other spits, and where mangroves have advanced seawards. Swan (1968) noted that some sectors had shown alternations of advance and retreat, and that locally the hinterland was being actively cliffed at high tide level while mangroves were spreading forward across pebbly mud in the inter-tidal zone. In general, the modern pattern of coastal change is related to the proximity of fluvial sediment sources, but previous phases of coastal advance occurred during the Holocene marine transgression when the rising sea swept weathered, wave-worked sands landwards to form beach ridges, and during a succeeding emergence which result in the construction of further sequences of beach ridges seaward of those previously formed (Swan, 1968). Over 90% of the east coast of Malaysia shows evidence of this Holocene progradation, but within the past century only a small part has continued to advance. Accordingly, this coast, facing the South China Sea, is one where, despite a substantial fluvial sediment yield, the trend from earlier progradation to more recent retreat of sandy coastlines is clearly discernible (Teh Tiong Sa, 1985).

EASTERN THAILAND

The extensive beaches along the shores of the Gulf of Siam are similar to those of the east coast of Malaysia: they are interrupted by minor rocky headlands and by lagoon and river mouths. At Pattani growth of the Laem Pho sand spit has protected the developing delta of Pattani River. In many sectors there has been long-term progradation, marked by successive beach ridges, particularly near river mouths, but locally the sandy coastline has retreated. Spits are growing at Laem Talumphuk, Laem Sui and Laem Yai, and delta growth has occurred at Mae Nam Ta Pi. Mangroves are advancing on sheltered bay shores east of Surat Thani and in Chumpon Bay (Pitman, 1985). The deltaic coastline at the head of the Gulf is prograding seaward at about 5 metres annually: sand bars are built up by wave action about 1.5 kilometres offshore to form a sheltered environment in which fluvial silt and clay settle, and as this land is formed it is reclaimed for agricultural use. South-east from Bangkok there are sectors of white sandy beach between headlands of granite and limestone at Pattaya and Rayong: exposures of beach rock indicate that there has been erosion in recent decades. East of Klaeng the coast is again swampy, with mangroves spreading on to tidal mudflats.

KAMPUCHEA

The embayed and island-fringed Kampuchean coastline includes steep and cliffed sectors exposed to the open sea, sandy beaches, especially to the rear of fringing reefs, and mangrove shores with some beaches and spits around the mouths of rivers such as the Klong Kâs Po, the Piphot and the Kos Sla. Apart from the seaward spread of mangroves on to accreting tidal mudflats, only minor changes have occurred during the past century.

VIETNAM

Sandy beaches alternate with cliffed promontories around the island of Phu Quoc, south of the Kampuchean border. Continuing progradation of the swampy deltaic coastline has been reported at the mouths of the Mekong River, as a sequel to recurrent major flooding in recent decades. Sand, silt and clay have been deposited at shoals around distributary mouths, and these have grown into swampy islands. Deposition has also advanced the fringes of the Mui Bai Bung peninsula to the south and west. The eastern coast, which has many steep sectors, some with receding cliffs, has changed little, except in the vicinity of river mouths where minor progradation has occurred on mangrove-fringed shores.

In the north-east there are some lengthy beaches, but the only substantial changes within recent decades are on the growing deltaic coastline around the mouths of the Song Hong River in the Gulf of Tonkin (Zenkovich, 1967). The southern shores of this delta have been advancing at up to 100 metres per year, the rate of progradation decreasing to the north-east where there has been some local erosion (Eisma, 1985).

CHINA AND HONG KONG

The southern coast of China, including Hainan, has many hilly promontories with minor cliffing and intervening bay beaches. Most of the changes during the past century have been caused by man's activities, continuing a long tradition of land reclamation and coastal modification for the construction of salt pans, fish ponds and harbour facilities, as well as coastal levees to prevent typhoon surges flooding low-lying farmland. Land reclamation has been extensive on the Zhujiang delta, the coastline of which has become a line of artificial dams enclosing fresh water areas, where vegetation is used to promote accretion prior to conversion into ricefields. Hailing Island, to the south-west, is one of several islands that have recently been linked to the mainland by causeways, alongside which sediment is accreting and mangroves are spreading.

The islands and territories of Hong Kong have a coastline of 738 kilometres, of which 246 are beach-fringed. So (1983) has reported that 22 kilometres advanced and 23 kilometres retreated during the past century, the remaining 693 kilometres having remained stable. Much of the coast is steep, and there are some retreating cliffed sectors, especially on weathered granites, but some sheltered embayments are occupied by mangroves, which have spread forward on to tidal mudflats. The main progradation has been at and around the mouth of Pearl River, which drains the South China province of Kwangtung. During the wet season (May through August) this river discharges large amounts of silt and sand into Deep Bay, and some of the sand has been sorted and distributed on to the adjacent shore, which faces north-west and is thus sheltered from strong wave action generated by the prevailing easterly winds, as well as from the summer typhoons, which come in from the east and south-east. The coastline is irregular in plan, with some prograded sectors projecting as sandy spits and forelands, and intervening parts cut back as cliffs in weathered granite, which yields quartzose sand to the nearby beaches. There has also been progradation of silty sand within a small embayment on the north coast of Lantau Island, which is sheltered by an offshore island. Parts of Hong Kong's beaches were temporarily swept away by the surge that accompanied Typhoon Wanda in 1962, and more recent changes have been documented by So (1981, 1983). Extensive land reclamation has taken place on the north coast of Hong Kong Island and the opposing Kowloon coast, and in some other filled embayments.

Little change has occurred along the generally rocky coast between Hong Kong and Shanghai, except for progradation near river mouths, notably on the Hanjiang delta, where much of the coastline consists of dams enclosing fresh water areas in course of reclamation. Chen Jiyu *et al* (1985) described the coastline of Central China as hilly and much indented, with sandy bay beaches rather than major spits or barriers, and low mangroves with salt marshes spreading on to tidal mudflats in sheltered areas. Hangzhou Bay has a large tide range (up to 8.9 metres): its southern shores are still prograding, whereas the northern shore has shown long-term recession, despite extensive construction of sea walls and groynes. The Changjiang estuary is an area of rapid shoaling and continuing accretion aided by land reclamation: it is subject to rapid changes during typhoon surges. To the north the Jiangsu coastal plain prograded rapidly when the Huanghe River opened to the sea here (between AD 1128 and 1855) but since the diversion of this river north into the Yellow Sea erosion has been extensive, attaining up to 130 metres per year around the abandoned river mouth. To the south, some of the eroded material is accreting on tidal mudflats sown with *Spartina* grass.

On the shores of the Yellow Sea, and especially in the Gulf of Bohai, there have been changes related to muddy sediment yield from the Huanghe, notably since the change in the mouth of this river back to its present position in 1855. Since then the deltaic coastline has prograded at up to 3 metres per year, about 2300 square kilometres of land having been added during the past century, and tidal mudflats now extend up to 10 kilometres offshore (Chen Jiyu *et al.*, 1984). Zenkovich (1967) mentioned sandy deposition in front of cliffs along the Shantung Peninsula, and on shores of the Gulf of Bohai there is active deposition of sand from the Luan to the north, as well as accumulation of shelly beach material. A sand spit 30 kilometres long, built down-drift from the mouth of the Luan, is backed by mudflats with an area of about 400 square kilometres, the new terrain having formed in less than a century.

TAIWAN

According to Hsu (1985) much of the western coast of Taiwan has been advancing as the result of accretion accompanied by tectonic uplift. In the Taipei area erosion predominates, despite tectonic uplift, while the coastlines of Yilan Bay in the north-east and from Tainan to Oluanpi in the south-west are also eroding. The east coast, generally steep, has shown little change during the past century.

KOREA

The western coasts of North and South Korea are generally low-lying, and strongly indented, with more than 2000 islands, and hilly promontories that are cliffed in the more exposed parts, where intertidal shore platforms have developed. Intervening bays have calcareous beaches. Tidal flats (600,000 ha) are extensive on the western and southern coasts, their area being maintained by accretion in front of reclaimed areas. Salt marshes are poorly developed because of the intensive reclamation: Dong Won Park reported that over 50,000 ha of tidal lands have been enclosed and reclaimed by South Korea since 1945, and the coastline has been shortened by up to 20% since 1910 by such enclosures. The eastern coasts of Korea are thought to be emergent: they are higher and steeper with many cliffed promontories, but where rivers deliver large quantities of sand and gravel there has been rapid beach accretion in embayments, particularly on Wonsan Bay, where spits and tombolos have been growing on a predominantly sandy coastline. Sand dunes are extensive along the east coast south from Wonsan Bay, some on spits and barriers enclosing lagoons, which are being silted by fluvial deposition. Similar accretion is in progress in Yeongil Man Bay, to the south, fed by the Hyongsan Gang River (Eisma and Park, 1985).

JAPAN

The total length of coastline of the four major islands of Japan is about 18,000 kilometres, much of it steep, with rocky and cliffed sectors. With a large population and intensive use of coastal regions, half the Japanese coastline is now either artificial as the result of reclamation, port and urban development (34%) or semi-natural, being fringed by roads and coast protection structures (16%). Sandy beaches make up about 16% (Koike, 1985). Comparison of surveys produced between 1890 and 1925 by the Geographical Survey Institute on a scale of 1 : 50,000 with maps produced by means of photogrammetry after 1945 by the same Institute on the same scale enabled

Koike (1977) to detect where sandy coastlines had advanced or retreated by more than 50 metres within this period. Ozasa (1977) traced the ensuing changes between 1946 and 1961. The following account is based on the findings of these two researchers.

On the north coast of the island of Hokkaido the sandy shores of spits and barriers have generally retreated, although some sectors of advance have been mapped. The south coast east of Tomakomai has prograded by up to 80 metres (1919–70) on shores that have been supplied with fluvial sediment from river catchments on the western slopes of the Hikada Range, but to the west the coastline has been cut back by up to 150 metres over a similar period. North-west of Sapporo the sandy coastline on either side of the mouth of the Ishikari River prograded 50–300 metres between 1916 and 1965, the gain being greater north-east of the river mouth, where one sector advanced 150 metres between 1947 and 1965.

Much of the Pacific coast of Honshu is cliffed, but there has been substantial progradation of sandy shores receiving fluvial sediment, notably from the Abukuma and Natori Rivers in Sendai Bay, or sediment derived from adjacent cliff erosion. The sandy coastline on either side of the mouth of the Abukuma prograded up to 100 metres between 1907 and 1968: essentially a very blunt delta built on a coast exposed to strong ocean swell. On the Shimokita Peninsula, where there are no large sand-yielding rivers, beaches have been cut back by up to 150 metres. On the Iwaki coast, south of Matsushima, cliffs up to 30 metres high in sedimentary rocks have retreated

Fig. 60 Construction of a sea wall on the coast of Byobugaura, Japan, to halt cliff recession. In the absence of basal wave attack, the cliff has started to degrade to a more gently sloping bluff. Photo: Eric Bird (August 1980)

at up to 0.7 metres per year. But there has been local progradation where northward-drifting sand has been intercepted by harbour breakwaters, as at Kashima. Cliff retreat at Byobugaura was about a metre per year until sea walls were built at the cliff base (Fig. 60). In the central part of Kujukuri-hama the sandy beach advanced by at least 200 metres between 1903 and 1967 (Fig. 61), the sand being derived from cliffs and beach erosion at both ends of the bay (Sunamura and Horikawa, 1977). Intensive reclamation and port development has made the coastline of Tokyo and Yokohama largely artificial, while successive earthquakes (e.g. in 1703 and 1923) in this region have caused minor uplifts, raising shore platforms above their level of form-ation, and triggered landslides, as in Sagami Bay.

South-west from Tokyo, in the Tokai District, the rivers that drain the Kiso and Akaishi Ranges (e.g. the Fuji, Abe, Oi and Tenryu) carry sands

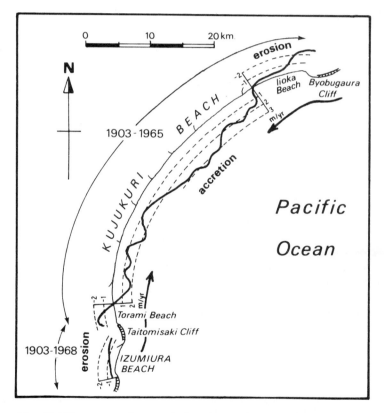

Fig. 61 Diagram to show variations in erosion and accretion rates on Kujukuri Beach, eastern Honshu, Japan, over the periods indi-cated. The accretion in the central sector is due primarily to conver-gence of longshore drifting of sediment eroded from the shores to north and south (Sunamura and Horikawa, 1977)

and gravels, and the gravelly deltas at their mouths have been prograding until the past 20 years, when erosion became prevalent. This is probably due to the construction of reservoirs and the extraction of gravel from river beds. At the mouth of the Tenryu the coastline retreated by up to 500 metres between 1917 and 1970, but shores to east and west showed progradation of up to 150 metres. On the Atsumi Peninsula, Yamanouchi (1977) measured cliff retreat of up to 43 centimetres per year.

On Shikoku there have been alterations of up to 500 metres advance and subsequent similar recession (1896–1969) on sandy shores fed by the Yoshino and Naka Rivers, and overall recession on the coastline of Tosa Bay. The east coast of Kyushu has shown historical alternations, but in Shibushi Bay the sandy coastline advanced about 150 metres between 1902 and 1965.

On the north coast of Kyushu there has been retreat of sandy shores but the east coast showed advance of up to 100 metres south of Takanabe between 1902 and 1966, with erosion of a similar amount farther south at Miyazaki. The north coast of Honshu, facing the Sea of Japan, has shown advance of the coastline at the head of Yumiga-hama of about 170 metres between 1899 and 1961. Elsewhere, retreat has predominated, necessitating anti-erosion works on the coast of Toyama Bay. Since about 1930, shores that were previously prograding near the mouths of the Shinano and Agano Rivers began to erode, partly because of the diversion of the mouth of the Shinano in 1922, and the subsequent growth of a delta at the new mouth, and partly because of reservoir construction. At Niigata the coastline retreated up to 400 metres between 1911 and 1968. The prevalence of beach erosion along the coast of Niigata Prefecture may also be due to increasing wave energy following the deepening of the offshore waters by current scour. The beaches of the Kanazawa coast generally retreated by up to 100 metres between 1909 and 1968, but there were localized advances near river mouths and to the north of the Kahoku-gata lagoon.

Koike (1977) calculated that of the sandy coastlines investigated, 21.3% had advanced and 17.2% retreated during the past 70 years. Comparisons of air photographs taken in 1946–8 and 1961 had shown gains on 14.3% and losses on 13.6% of these coastlines (Tanaka, 1973, 1974). It was deduced that erosion had become more prevalent in recent years as a consequence of diminished fluvial sediment yields following reservoir construction, channel diversion, and the extraction of sand and gravel from river beds. Taking this into account it seems that Japan, with its steep coasts, high catchment runoff and sediment yields and orogenic instability, was previously (like New Zealand) a region where sandy coastlines were naturally maintained or prograded by deposition.

Cliff erosion has been widespread in Japan, but many cliffed sectors are now protected by sea walls and heaps of tetrapods (Fig. 62). The extent of erosion has led to widespread dumping of large concrete tetrapods on foreshores, producing an unsightly coastline, inhospitable as a milieu for beach and nearshore recreation. Active vulcanicity has led, directly or indirectly,

Fig. 62 Fifty-tonne concrete tetrapods have been dumped on the shore at Atami, near Tokyo, Japan, to protect the waterfront against waves generated in typhoons. The formerly beach-fringed coastline has thus been artificially stabilized Photo: Eric Bird (August 1980)

to progradation on parts of the Shiretoko Peninsula in northern Hokkaido, subsequently cliffed by marine erosion, and has tied Sakura-jima island to the mainland in Kagoshima Bay. In 1973–4 successive eruptions built a new volcanic island, Nishino-shima Shinto, in the Izu-Ogasawara arc, south of Japan, an island similar in origin to Surtsey, off Iceland.

PACIFIC USSR

The Pacific coast of the Soviet Union includes long sectors of steep and mountainous coast, with receding cliffs up to 700 metres high in eastern Kamchatka and northward along the shores of the Bering Sea. On the lower western shores of Kamchatka peninsula there has been rapid recession of cliffs of sandy loam, especially during storms (Zenkovich, 1967). There have, however, been sectors of major deposition during the Holocene, notably in north-eastern Sakhalin, where Zenkovich described sand and gravel barriers up to 5 kilometres wide and more than 300 kilometres long in front of coastal lagoons, with only limited stretches of active cliffing in friable Tertiary deposits. The patterns of beach ridges on the barriers have been used to decipher a complicated history, but it is evident that much of the sediment was carried shoreward from the sea floor during, and to some extent since, the major Holocene marine transgression. At the present time this barrier coastline is being cut back, with erosional processes removing beach ridge terrain that was built up previously. This is consistent with the sequence of events on barrier coastlines elsewhere, notably in Australia. Y. D. Shuisky calculated that 56% of the coastline of Sakhalin was retreating at rates of up to 5 metres per year; 34% was stable, and 10% advancing, locally up to 10 metres per year in river mouth depositional areas.

Vulcanicity has locally influenced coastlines, for example in the Sea of Okhotsk, where a new volcanic island (Taketomy) was built in 1933 and then eroded into cliffs of ash and scoria, yielding material that was deposited in trailing spits that grew out towards the adjacent coast of Alaid Island, forming a double tombolo (Zenkovich, 1967). Deltaic progradation has continued at the mouths of the Okhota and Kukhtuy Rivers, and the spit near the mouth of the Moroshechnaya River in north-west Kamchatka has also grown. The features of easternmost Siberia are steep and mountainous, with some barrier-enclosed lagoons, but the coastline is retreating, and sectors of active progradation have been limited to a few river deltas (Kaplin, 1985).

ARCTIC USSR

The Arctic cost of the Soviet Union has features comparable with those of Arctic Canada, with a similar sea-ice regime. Progradation of marshes and mudflats has continued near river mouths, but the Lema delta is being eroded. Y. D. Shuisky reported that cliff retreat is in progress on 45% of the Chukchi Sea coastline, 10% of the Laptevykh Sea coastline, and 30% of the Barents Sea coastline. Tundra cliffs are similar to those of northern Canada, subject to freeze-thaw disintegration as well as to erosion by summer wave action, and Zenkovich (1967) quoted records of cliff recession in excess of 50 and even 70 metres in a summer, with complete destruction of outlying islands such as Semenovskii and Vassilevskii. Typically, average cliff recession is 4 to 6 metres per summer, with up to 15 metres on Morzhovyets Island. Locally, driftwood mobilized by waves has accelerated cliff erosion. In sheltered situations (embayments behind barriers) tidal flats are being built from clays eroded and transported away from these cliffs. Erosion is slow on hard rock outcrops on Kola Peninsula, Vaigatch Island, Novaya Zemlya, the northern Taymyr Peninsula, and Wrangel Island. Sectors of beach progradation (e.g. the large spit at Sengeyski on the Barents Sea coast, and the barrier at Billings Cape on the Chukchi Sea coast) are related to sand and gravel supply from eroding tundra bluffs rather than from rivers, which yield mainly fine-grained sediment.

PHILIPPINES

The Philippine Islands are generally hilly or mountainous, with steep coasts locally cliffed, and alluvial lowlands including some prograding deltas and beach ridge plains. There are active volcanoes, some of which have delivered pyroclastic material directly to prograding coastlines while others have augmented fluvial sediment loads delivered to river mouths. Earthquakes are common, and have locally modified coastlines, especially where they caused landslides in deeply weathered rock materials on steep coasts. Mangroves are extensive in sheltered inlets and embayments, and in some places are advancing on to depositional tidal flats; but there has been extensive embanking of mangrove areas for the construction of fish ponds, and some swampy coastlines have become artificial as a consequence of this (Bird, 1985).

On the north coast of Luzon there has been depositional progradation at and around the mouth of the Cayagan River, with gains on bordering beachridge plains. The west coast is steep, and locally cliffed: at Luna an eighteenth-century coastal watchtower has recently been undercut by marine erosion. In Lingayan Gulf the Bauang and Aringay Rivers have built small lobate deltas, and southward drifting of beach material has extended the spit at Caboroan. Fluvial deposition has also prograded the beach ridge plain on the southern shores of this gulf. On the volcanic Bataan Peninsula cliffs have been cut back on interfluvial promontories, and in adjacent Manila Bay there has been rapid growth of the large delta built by the Pasag, Pampanga and Bulacan Rivers, the prograding swampy coast being converted quickly into fishponds (Fig. 63). South of Manila there has been continuing accretion on the Cavite spit. Southern Luzon has many inlets with mangroves, but the more exposed Pacific coast has sandy beaches between cliffed promontories. Near Legaspi the black sand beaches have been augmented by volcanic material washed down from the Mayon volcano, while in Larap Bay sedimentary waste from iron ore quarrying has flowed down rivers to prograde the coastline.

Fig. 63 Rapidly growing delta on the coast north of Manila, Philippines, where the newly accreted land has been converted into brackish water fishponds. Photo: Eric Bird (July 1969)

Mindoro is a hilly to mountainous island with fringing reefs, beaches and swamps. Changes have taken place in recent decades on the growing Mongpong, Caguray and Saquisi River deltas, and sand from the Bagsanga River has drifted southwards to prograde beaches and enlarge the Caminawit spit. There has been minor retreat on cliffed promontories.

Palawan is steep and reef-fringed, with retreating limestone cliffs and beaches of coralline sand. Panay, similar to Mindoro, has several growing river deltas, notably the Iloilo and Jalauod in the south-east and the Panay and Aklan in the north. Sand from the Sibalom River has drifted south to prograde the foreland at Tubingan Point, near San Jose.

On Negros Island the Malogo and Bago Rivers have prograded their deltas and adjacent beach ridge plains. Cabu Island has slowly retreating cliffs in emerged coral limestone. Otherwise steep coasts predominate in the southern Philippines, with local progradation near river mouths. On the east coast of Leyte, sand drifting northwards has built up the spit at Cataisan Point.

On Camiguin Island, north of Mindanao, the active Hibok volcano has prograded the coastline by lava and ash deposition from successive eruptions. On Mindanao, sediments from the Agusan River have prograded beach ridge plains in Butuan Bay, and there has been similar progradation in Macajalar Bay. In the south-east, landslides are common on steep coastal slopes, and there are high retreating cliffs on Cape San Agustin.

INDONESIA

Coastline changes in Indonesia have been reviewed by Bird and Ongkosongo (1980). This archipelago of about 13,700 islands with a total coastline exceeding 60,000 kilometres, has shown rapid changes on deltas, in mangrove swamps, in areas affected directly or indirectly by volcanic activity, and on some retreating cliffs, but there are extensive steep coasts, with little or no basal cliffing, where changes have been very slow.

In northern Sumatra formerly prograded beach ridge plains have been truncated by erosion at Padang, Sirbangis and Kutajara. Verstappen (1973) described deltas with prograding coastlines, notably the new delta built by the Peusangan River after a change in its course led to the abandonment and erosion of an adjacent older delta. Beach ridge plains occur at Cape Intem, on the southern shores of Malacca Strait, and the Singkel plain in south-west Sumatra, but it is not known if progradation has continued in these areas. The swampy north-east coast of Sumatra has certainly prograded in recent centuries by the deposition of fluvial sediment (as on the Batang Hari delta) and the advance of mangrove swamps. Palembang was probably a river-mouth port when Marco Polo visited it in 1292, but it is now 50 kilometres upstream behind a swampy coastal plain. Patterns of active erosion and deposition alongside estuaries north of Palembang have been mapped by Chambers and Sobur (1975), and are due partly to channel meandering and partly to mangrove encroachment on to tidal mudflats.

Cliffed coasts are found on the south-west coast of Sumatra and its outlying islands, and also along the south coast of Java and the islands eastwards to Sumba.

Vulcanicity has contributed to coastline changes locally in Indonesia during the past century, most notably at Krakatau (Fig. 64), a volcanic island in Sunda Strait, where a major explosion in 1883 completely reshaped the topography, leaving a ring of islands of pyroclastic material, on the shores of which cliff erosion has subsequently been very rapid: up to 40 metres per year. In 1927 a new volcanic island, Anak Krakatau (Fig. 65), appeared in the waters enclosed by the residual islands, and this has since grown by

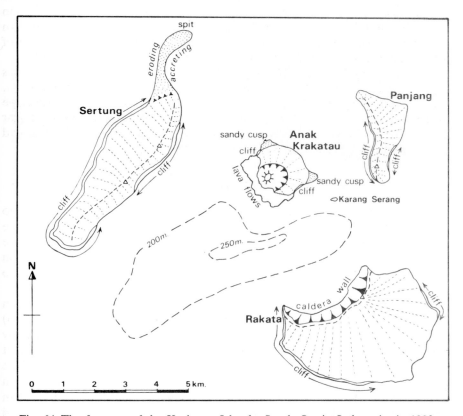

Fig. 64 The features of the Krakatau Islands, Sunda Strait, Indonesia, in 1983, a century after an explosive eruption destroyed much of a large volcanic island on this site. Sertung, Panjang and Rakata were three ash-mantled residual islands left after the explosion, and cliffs cut in volcanic ash have receded, especially on their more exposed western and southern coasts. In 1927 Anak Krakatau appeared in the centre, and became enlarged by successive eruptions of ash, and later of lava. Its south-western coasts have a bastion of solid lava, but cliffs are being cut back in volcanic ash, and derived sand and gravel have accumulated on two cuspate forelands and an intervening beach on the northern shore

128

Fig. 65 A view of Anak Krakatau, a volcanic island that has grown up since 1927 within the caldera of the former Krakatau volcano. The irregular western coast consists of lava flows (L), while to the east there are rapidly retreating cliffs (C), cut in volcanic ash deposits, and sediment derived from these has accumulated in a sandy cuspate foreland (F) at the eastern end of the island. Photo: Neville Rosengren (June 1981)

successive eruptions of lava and ash, its coastlines changing rapidly as the result of depositional progradation followed by marine cliffing, with recession at up to 6.6 metres per year, and lee-shore spit growth (Bird and Rosengren, 1984).

The beach ridge plains of West Java have shown cliff and beach recession in the past few decades, as a sequel to the damage caused by the tsunami that resulted from the 1883 explosion of Krakatau. On the north coast, comparisons of maps made at various times since 1865 with modern air photographs show rapid and intricate gains and losses along the deltaic coastlines: gains at and around river mouths, losses on abandoned deltaic lobes. In some cases, deltas have grown out from artificial canals cut to divert river outflow: on the Cidurian delta the coastline advanced up to 2.5 kilometres seaward alongside such a canal between 1927 and 1945, by which time the earlier delta had been much eroded. Changes on the shores of

Jakarta Bay between 1869 and 1940 resulted in the addition of 26 square kilometres of land, the very rapid accretion probably resulting from hinterland soil erosion (Verstappen, 1953). The northern part of the Citarum delta prograded by up to 3 kilometres between 1873 and 1938, but the western part, facing Jakarta Bay, was cut back by up to 140 metres. Accretion slowed after the completion of the Jatiluhur Dam in 1970, whereafter the fluvial sediment supply diminished. Changes on the Cipunegara delta, 1865 to 1978, included the building and destruction of one delta lobe, and its replacement by another lobe to the north, and similar changes have been recorded on the Cimanuk delta (Fig. 66) between 1857 and 1974 (Hehanussa, 1979), including the growth of a new branching delta north-east of Indramayu, after a new north-eastern outlet was produced by a major flood in 1947 (Fig. 67). Bird and Ongkosongo (1980) have mapped the patterns of change on the

Fig. 66 Pancer Payang, a distributary of the Cimanuk River, north Java, which has built its delta rapidly in recent years by deposition of sediment from successive river floods. As soon as the new land is deposited, embankments are built on it to enclose fishoponds. Photo: Neville Rosengren (June 1981)

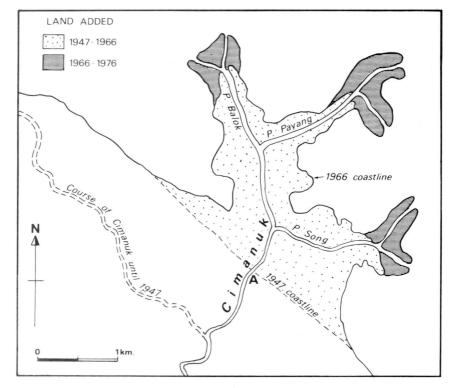

LAND ADDED

1947-1966

1966-1976

P. Balok

P. Payang

1966 coastline

Course of Cimanuk until 1947

N

Cimanuk

P. Song

A

1947 coastline

0   1km.

Fig. 67 Diverted to a new north-eastern outlet during a major flood in 1947, the Cimanuk River has built a substantial modern delta, with rapid deposition along-side branching distributaries. The earlier delta to the west, no longer receiving fluvial sediment, is being cut back by marine erosion (after Hehanussa, 1979)

Bangkaderes, Sanggarung, Bosok, Pemali, Comal, Bodri and Solo deltas (see also Höllerwoger, 1966 and Verstappen 1977), and the large-scale progradation on the coastline south of Japara, up to 5 kilometres between 1911 and 1972, alongside the Wulan Canal. Rapid modern progradation is thought to be due to soil erosion resulting from deforestation and the extension of cultivation in hilly hinterlands.

North Java also has steep coasts, and mangrove-fringed areas, particularly in embayments. Outlying coral islands and cays off Jakarta have shown intricate changes in configuration, traced by Zaneveld and Vestappen (1952) from maps made in 1875, 1927 and 1950. The south coast of Java, exposed to stronger wave action, is generally cliffed or beach-fringed. Eruptions of Merapi volcano, north of Jogjakarta, have augmented fluvial sediment yields and supplied sand to beaches west from Parangtritis, but the extent of progradation during the past century is not known. In the rapidly silting Segara Anakan embayment, mangroves have advanced several kilometres in

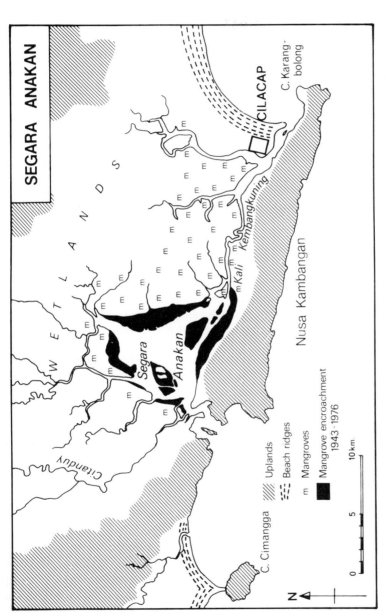

Fig. 68 Rapid advance of mangroves has taken place in recent decades on accreting mudflats within the shallow estuarine lagoon of Segara Anakan, in southern Java. Accretion has been rapid, largely because of heavy sedimentation from the Citanduy River, which drains a catchment within which there has been extensive soil erosion

recent decades (Fig. 68). Farther west, O.S.R. Ongkonsongo has reported beach erosion.

The islands to the east of Java have generally steep coasts, often fringed by coral reefs or mangrove swamps, and it is unlikely that there has been much change here during the past century, except on cliffed southern coasts and where deposition has occurred around river mouths. The southern shores of Kalimantan are deltaic and swampy, and some parts are certainly advancing. On the east coast the Mahakam delta is growing, but otherwise there has been little change. The same is true on the steep, indented coasts of Sulawesi, but in Irian Jaya there are prograding swampy shores, particularly in the south-west, where the swampy Cape Valsch coastline has advanced rapidly. Tectonic movements have modified coastal terrain, for example on the Mamberamo delta on the north coast, where subsidence created lakes, caused a regression of mangroves into fresh water swamps, and diminished fluvial sediment supply to the river mouth where erosion ensued. Earthquakes have triggered coastal landslides in northern Irian Jaya and on outlying islands, forming temporary progradational lobes.

PAPUA NEW GUINEA

There is widespread evidence of Holocene tectonic activity in Papua New Guinea, and earthquakes continue, their chief effect being to augment fluvial sediment yields and thus promote coastline progradation. The 1907 earthquake in the Torricelli Ranges on the north coast had this effect, and also caused subsidence of part of the coastal plain west of Aitape to form Sissano Lagoon. The 1935 earthquake in this area also produced extensive landslides, and increased sediment yields to river mouths. Sand and gravel drifting eastward have accumulated on a prograding beach at Cape Wom, but beaches at Wewak, deprived of this supply by coastal uplift, have been diminished by attrition and are now eroding (Bird, 1981b). Much of the north coast is steep, with occasional landslides but only minor and localized cliffing.

Progradation has continued at the mouth of the Sepik River but the sandy northern shore of the delta is subsiding and retreating, the beach migrating landward over submerging mangrove swamps and lagoons produced by subsidence. Coastal subsidence also appears to have prevented progradation of the Ramu delta, to the east. Vulcanicity has built and enlarged islands offshore between Wewak and Madang, and at several places around the steep coasts of New Britain, notably at Rabaul. The north coast of Huon Peninsula has been intermittently uplifted by tectonic movements, with the result that the coastline has advanced up to several kilometres in the late Pleistocene, but no advance has been reported within the past century (Chappell, 1974). The emerged coral coasts are being cut back as low cliffs with characteristic 'notch-and-visor' profiles (Löffler, 1977).

Coastline changes have been minor around Huon Gulf and on the coast

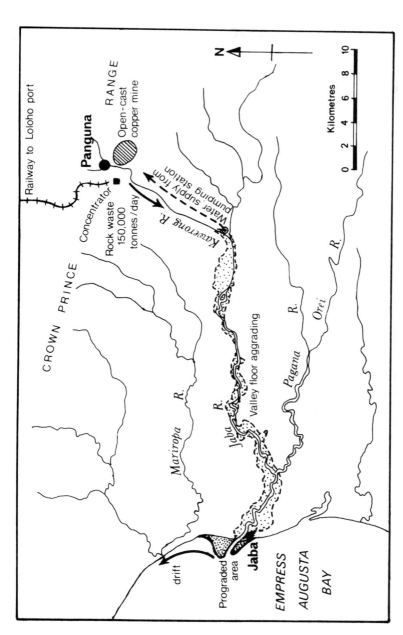

Fig. 69 Sediment flowing down the Jaba River on Bougainville, northern New Guinea, has been augmented in recent decades by rock waste from the Panguna copper mine in its headwaters. As a consequence, there has been deltaic progradation at the river mouth, and beach accretion on the shore of Empress Augusta Bay to the north (after Brown, 1974)

of Papua. The Mount Lamington coastline was prograded by landslides, mudflows, and deposition of sediment washed down rivers during and after the 1951 eruption. On the south coast there are receding cliffs on promontories, especially behind gaps in the barrier reef, as between Samarai and Mullins Harbour, where wave energy is stronger. Sediment supplied from rivers has been built into prograding beach ridge plains, as in Hood Bay, where sand arrives from the Kemp Welch River, but the cuspate foreland at Kwaipomata Point has eroding flanks. At Iokea, erosion of the southern side of a lobate beach ridge plain has been compensated by accretion to the north, and Auma Spit has been built and extended east of the mouth of the Vailala River. Local spit growth has occurred at many places, notably at Kerema.

Sand supplied from distributaries of the Purari River forms shoals, beaches and longshore spits that have grown in front of mangrove swamps on the swampy coast of the Gulf of Papua. Changes at the mouth of the Ivo-Urika River between 1956 and 1973 included the enlargement of shoals and mangrove islands, especially in the lee of an accreting river-mouth bar, while between river mouths the delta margin is eroding, with mangrove trees undercut by wave attack, as to the west of the mouth of Varoi River (Thom and Wright, 1983).

Bougainville is a steep, mainly volcanic island, with beaches, spits and deltas supplied with sand and gravel derived from the erosion of volcanoes, and augmented by volcanic eruptions. In the south-west the Laruma, Tarokina and Saua Rivers have had their loads periodically increased by volcanic material from Mount Bagana, with ensuing beach progradation, but between eruptions these beaches have been eroded. Moila Point is a major cuspate foreland that has prograded in the lee of Shortland Island to the south. Open-cast copper mines at Panguna have generated vast quantities of waste material which have increased the sediment load of Kawerong River and resulted in delta growth at Jaba (Fig. 69) and progradation of beaches along the adjacent shores of Empress Augusta Bay (Brown 1974).

AUSTRALIA

There has been local and minor progradation of low-lying coastlines in the Gulf of Carpentaria during the past century, but the early charts are not sufficiently accurate to determine the extent of such changes. The Arnhemland coast includes steep and cliffed sectors as well as beaches and spits, but the southern shores of Van Diemens Gulf are low-lying and swampy, and local progradation has taken place near river mouths and on cuspate forelands such as Point Stuart (Fig. 70). It is also probable that mangrove swamps, extensive in north coast inlets (e.g. Port Darwin), have advanced locally during the past century. On Bathurst Island there has been retreat of gullied cliffs cut in laterites (Galloway, 1985).

Fig. 70 Successively formed ridges of coralline sand and gravel on a prograding coastline at Point Stuart, Northern Territory, Australia. Photo: Eric Bird (May 1970)

Joseph Bonaparte Gulf is fed by the Victoria and Ord rivers, and bordered by extensive prograding depositional flats, but the rocky shores of the Kimberley coast have changed very little, although mangrove encroachment on to accreting mud flats has advanced some sectors of King Sound while others have been cut back (Jennings, 1975). On the north coast sand has moved in from the sea floor to prograde beaches at Cape Leveque, and there has been some accretion on spits bordering tidal inlets, as at Broome. The sandy Eighty Mile Beach has extensive sectors with cliffed backshore dunes indicating recent recession, but intermittent accretion has occurred on the coastline of the De Grey delta as a sequel to episodes of river flooding. Beach erosion near Port Hedland has exposed beach rock, and there has been slight recession of low cliffs cut into dune calcarenite. Farther west, in the Onslow district, there have been minor changes on the sandy barrier islands and spits, interesected by tidal inlets, in recent decades, the major changes occurring during cyclones, which bring river floods and sea surges. At Onslow there has been sand accretion alongside the harbour breakwaters.

Much of the west coast of Australia is bordered by dune calcarenite formations, extensively cliffed. Beaches occur intermittently, and there has been local progradation, notably near the mouths of rivers, such as the Gascoyne near Carnarvon (Lustig, 1977). At Geraldton there has been retreat of up to 40 metres since 1880 on Horrocks Beach, but in general the west coast beaches have remained stable, with local progradation north and south of Denison, at Jurien, south of Cervantes, Wedge Island and Pinaroo Point, in each case due to shoreward movement of sea floor sand. Several sectors of dune calcarenite have receding cliffs, and beach erosion has occurred at Cottesloe. At Scarborough Beach, north of Cottesloe, Eliot *et al.* (1982) measured seasonal cycles of advance and retreat of up to 25 metres, superimposed on a steadier progradation of about 3 metres per year on this sandy coastline. Artificial structures and local reclamation have modified the Perth coastline and the shores of Cockburn Sound. Alternations of erosion and deposition have changed the coastline around Mandurah, and near Bouvard Reef a sandy beach has advanced up to 60 metres in front of a former cliffed dune. The Bunbury coastline has shown alternations with little net change, but north of Wonnerup there has been recession, with the foundations of early European settlement sites being exposed. To the south, there have been gains of up to 4 metres on some sectors since 1850, and similar losses on others.

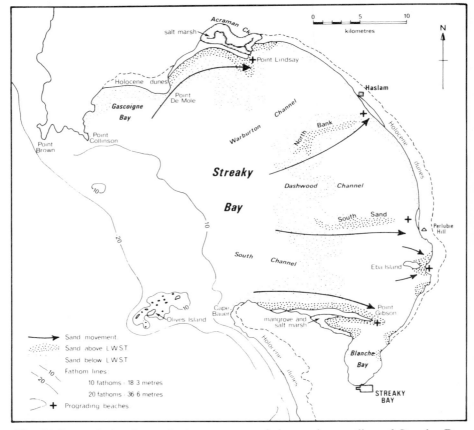

Fig. 71 Progradation is continuing on parts of the sandy coastline of Streaky Bay, South Australia, inshore from elongated shoals, whence sand has moved shoreward to built up flanking spits (Point Lindsay, Point Gibson) and cuspate forelands

On the south coast of Western Australia, bay beaches between rocky promontories are either stable or retreating. Hodgkin (1976) has traced short-term changes from maps made in 1832, 1878, 1920, 1925–29 and air photographs taken in 1943 and 1973 on sandy coastlines at the mouth of the Blackwood River, near Augusta, but no long-term erosion or progradation could be demonstrated. Similar changes have occurred around several lagoon entrances here. Beach erosion has been substantial in the southern part of Esperance Bay, but W. Andrew reported that the northern part has shown accretion in recent decades.

The coastline bordering the Nullarbor Plain is one of retreating vertical, crenulate cliffs cut in soft limestones. Jennings (1963) quoted evidence of sandy coastline advance at Twilight Cove (Fig. 3), indicated by a nineteenth-century shipwreck which is now some 200 metres inland. The intricate peninsulas on the west coast of the Eyre Peninsula are mainly of dune calcarenite, cliffed where they are exposed to the ocean, with some harder outcrops of granitic basement, on which erosion has been extremely slow. Streaky Bay (Fig. 71) is notable for its extensive shoals, from which sand and shelly material have moved onshore to prolong spits and prograde cuspate forelands during the past century. Spits have also some grown slightly in recent decades in Coffins Bay, to the south.

Minor changes have occurred on the shores of Spencer Gulf and Gulf St Vincent, both on retreating cliffs and where deposition has continued to prograde spits and cuspate forelands. Mangrove encroachment has proceeded

Fig. 72 Ocean swell moves shelly sand on to the shore of Encounter Bay, South Australia, but there has been little progradation in recent decades because much of the accreted sand is blown landward in parabolic dune formations. Photo: Eric Bird (March 1979)

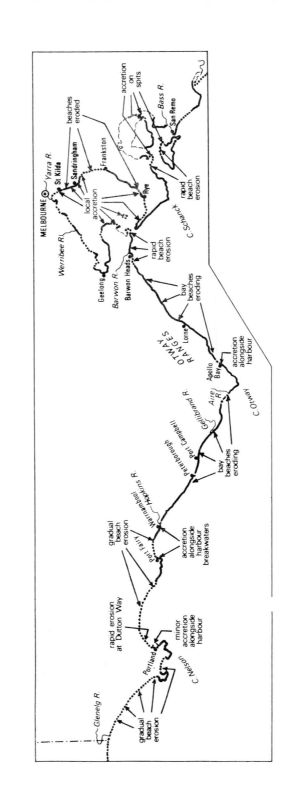

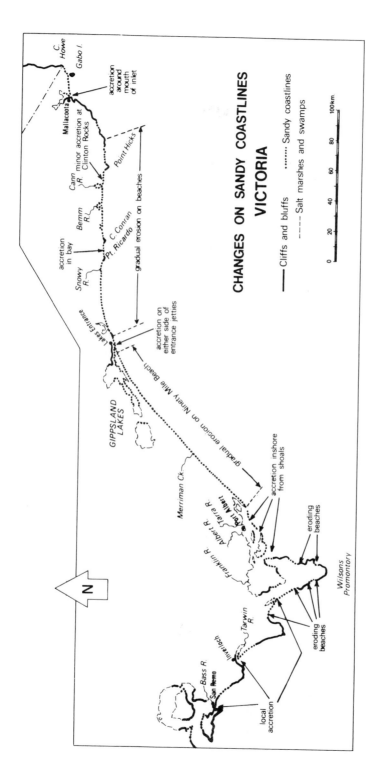

Fig. 73 Evidence from historical maps, charts and air photographs has been used to compile this map showing changes on sandy coastlines in Victoria, Australia

locally in bordering inlets, and near the northern ends of these gulfs. At Adelaide, northward drift of beach sand has resulted in large-scale accretion at Glenelg, and at Largs alongside the breakwater at Outer Harbour. To the south, progradation of up to 100 metres has resulted from the spilling of gravelly quarry waste on to the shore at Rapid Bay, where the beach advanced up to 100 metres between 1945 and 1975.

Near Victor Harbour there has been progradation of up to 15 metres during the past half-century at the mouth of the Inman River, and erosion at the mouth of the Hindmarsh River. Bourman (1974) suggested that soil erosion in the catchment of the Inman had augmented its sand yield, in contrast with that from the less eroded Hindmarsh basin. He also noted the rapid erosion at Middleton, where a broad beach backed by dunes has been removed since the beginning of the century, and alluvial terrain to the rear is now being cut back in cliffs; a sequence possibly initiated by local subsidence during a 1902 earthquake. Minor alternations have occurred at the Murray mouth since 1839, successive maps and air photographs showing slight migrations in its position, and episodes of widening and narrowing (Harvey, 1983).

The long sandy and shelly beach bordering Encounter Bay (Fig. 72) has changed little during the past century, but retreat has proceeded on other sandy sectors in the south-east of South Australia, and along much of the Victorian coastline (Fig. 73). In Victoria, cliffed coasts are dominant on the ocean shore, and progradation has been confined to beaches adjacent to harbour breakwaters, as at Port Fairy, Warrnambool, Apollo Bay and Queenscliff, and to the ends of growing spits, as at Swan Island in Port Phillip Bay, and Sandy Point and Observation Point in Westernport Bay. Beach erosion has been rapid near Portland, at Lorne, Point Lonsdale, and Cowes, and extensive sea walls have been constructed, especially in Port Phillip Bay (Fig. 74) (Bird, 1980). Erosion has developed on beaches in Cleeland Bight, on Phillip Island, and on the Yanakie isthmus, Wilsons Promontory, after the supply of sand to the shore from spilling dunes was reduced by the planting of marram grass to halt drifting sand (Fig. 75). Accretion has continued, however, on barrier islands east of Corner Inlet, where sand is moving in from extensive shoals offshore. The Ninety Mile Beach has generally been cut back in recent decades, but there has been local accretion on either side of harbour breakwaters (Fig. 76) built at Lakes Entrance in the late nineteenth century (Fig. 77). However, between 1957 and 1982 there was little progradation here, the sand arriving on the beach being swept by onshore winds to heighten dunes developing above high tide mark. At Gabo Island a tombolo has formed intermittently during the past century (Ballard, 1983). Minor changes have been detected on mangrove-fringed shores in Westernport Bay, especially where the mangroves have disappeared, partly as a consequence of man's activities, and erosion has ensued (Bird and Barson, 1975).

In Tasmania, progradation has been confined to minor growth on spit

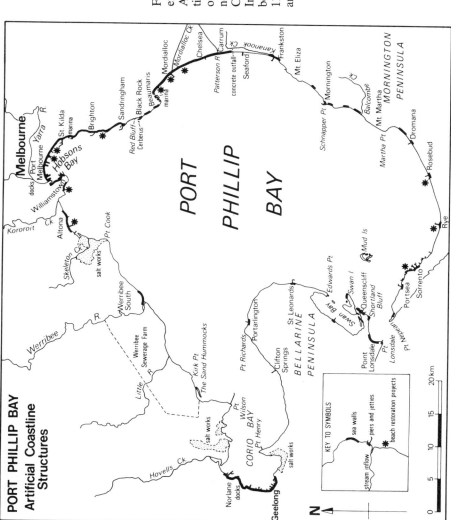

Fig. 74 The response to coastal erosion around Port Phillip Bay, Australia, has been the construction of sea walls, now extensive on the urbanized coast to the north-east (Melbourne to Carrum) and in the Geelong area. In recent years, artificial sandy beaches have been introduced at 11 sites to aid coastal stabilization and restore a recreational resource

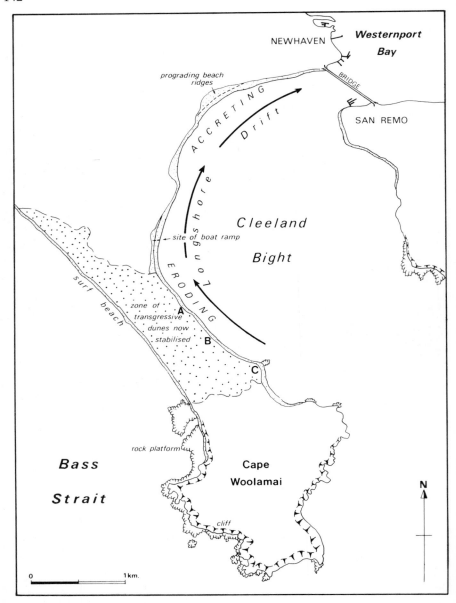

Fig. 75 In recent decades dunes drifting across the sandy isthmus at Cape Wool-amai, Victoria, spilled on to the shore of Cleeland Bight (at A, B, C), augmenting the beach. Since 1975 the dunes have become stable as the result of successful planting of grasses and shrubs by the Soil Conservation Authority, and the supply of sand to the southern shore of Cleeland Bight has ceased. As a result, the beach has become depleted, longshore drift carrying sand northward to an accreting beach near Newhaven

Fig. 76 Sandy forelands (A, B) produced by accretion on either side of the break-waters built in 1889 at Lakes Entrance, Victoria, Australia (cf. Fig. 77) Photo: Neville Rosengren (October 1980)

formations at Weymouth, Rheban Spit and Marion Bay, and to accretion alongside groynes and breakwaters, as at Devonport, Ulverstone, and the entrance to Macquarie Harbour. The southern end of Ocean Beach near Strahan has prograded slightly, while on Seven Mile Beach near Hobart, erosion at either end has been balanced by progradation of the central part. Elsewhere, the Tasmanian coast has remained stable, or has receded slightly during the past century (Davies, 1985). Beach erosion has been prevalent on the sandy shores of King Island and Flinders Island during the past century, but on the east coast of Flinders Island there has been accretion on the cuspate foreland at Sellar Point.

On the New South Wales coast there has been minor recession on cliffed sectors, and on sandy coastlines in embayments. Progradation has been generally restricted to the vicinity of artificial structures, such as the harbour breakwaters of Moruya and Tweed Heads (Thom, 1974), but natural

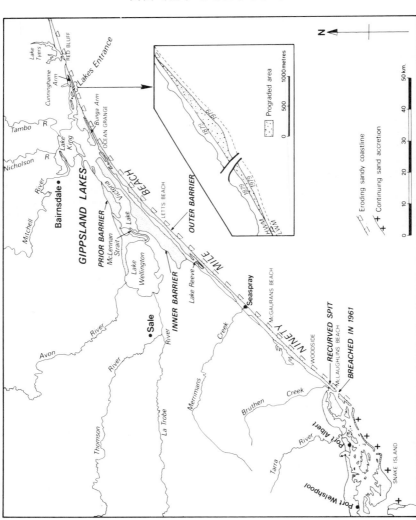

Fig. 77 Changes on the Ninety Mile Beach in south-eastern Australia during the past century have included coastline recession, with cliffs cut into backshore dunes, except on the sectors on either side of the breakwaters built at Lakes Entrance in 1889, where there has been sand accretion. Local progradation of beaches on the barrier islands to the south-west has resulted from shoreward drifting of sand from shoals

progradation on Seven Mile Beach has been related to sediment supply from the adjacent Shoalhaven River (Chapman *et al.*, 1982). While many beaches show evidence of long-term recession, others have maintained their alignments. Bryant (1983 a, b) used numerous ground and air photographs (1890–1980) to show that the beach at Stanwell Park, south of Sydney, has shown phases of accretion and of erosion, evidently related to regional or global climatic variations rather than local storminess: over the 90-year period there was no distinct trend, either of erosion or accretion, on this beach.

On the east coast of Queensland there are alternating sectors of steep coast with minor cliffing, and sandy beaches, which are generally receding, notably along the ocean shores of North and South Stradbroke Island, Moreton Island, and Fraser Island. Beach erosion has become a major problem at resorts such as the Gold Coast south-east of Brisbane, where extensive walling and artificial beach renourishment have been introduced. Sand drifting northwards has been intercepted by the breakwaters at Tweed Heads and Kirra to give local accretion. Measurements of changes have been documented by the Queensland Beach Protection Authority at a number of localities, for instance at Flying Fish Point, near Innisfail, where there are records of sandy coastline progradation by up to 100 metres between 1883 and 1922, followed by intermittent recession of up to 90 metres between 1922 and 1968 (Macdonald *et al.*, 1973).

Fig. 78 Sand brought down by the Barron River, North Queensland, Australia, during floods has been deposited off the river mouth (B) and subsequently washed up by waves on to adjacent beaches. In the middle distance the coastline has prograded about 300 metres between 1942 and 1960, and a spit is growing south-wards in the distance at Casuarine Point (C). Photo: Eric Bird (Agust (1969)

146

Offshore, there has been a predominance of erosion on the sandy shores of cays in the Great Barrier Reef area in recent decades, beach rock being extensively exposed. Yet on some of these islands the supply of reefal sediments to beaches may have increased following recurrent depredations of coral by the crown-of-thorns seastar (Stoddart *et al.*, 1978).

In North Queensland, fluvial sediment yields have been a major source of local progradation at and near river mouths, for example on the Burdekin delta, where Pringle (1983) has traced changes on the coastline between 1942 and 1980, including the northward growth and migration of spits and barrier

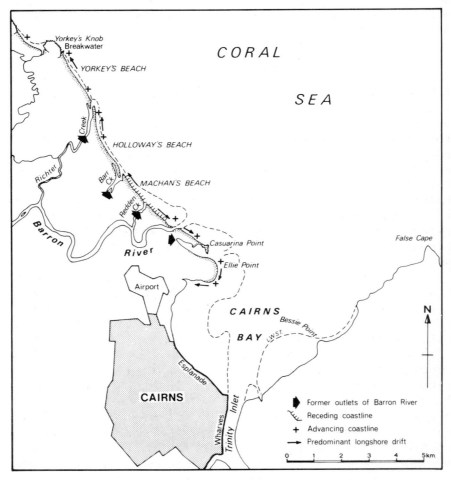

Fig. 79 Patterns of coastline change on the Barron Delta, North Queensland, have been related to waxing and waning of river distributaries. A former distributary lobe at Machan's Beach has been cut back by erosion, some of the sediment drifting northward to built up Holloway's Beach, and some southward to nourish a growing spit at Casuarina Point

islands by up to 3 kilometres. Changes have proceeded on a smaller scale on the Barron delta (Fig. 78) near Cairns (Bird, 1969). In such cases there has been progradation on spits and cuspate forelands, but even in deltaic areas there are sectors of active erosion (Fig. 79). Mangrove-fringed shores within embayments such as Princess Charlotte Bay may have advanced seaward during the past century, but early surveys are not sufficiently accurate for this to be measured.

Recession has thus been prevalent on the Australian coast during the past century, including sandy barriers that had previously prograded, with the addition of successive beach ridges earlier in Holocene times. The onset of erosion occurred at various dates within the past 6000 years (Thom, 1978), and there are still a few places where progradation is continuing.

NEW ZEALAND

McLean (1978) described the coastline progradation that has occurred in various parts of New Zealand during the past century. Fluvial sediment yield, augmented by the effects of tectonic and volcanic activity, as well as by deforestation and cultivation in steep catchments, has been substantial enough to maintain or prograde sand and gravel beaches in Hawke Bay on the east coast of North Island, and on the west coast of Wellington Province, where the sandy beaches have advanced up to 2 metres per year. At Santoft Beach the coast has advanced 100 metres since the *Fusilier* was wrecked there in 1884. Drifting sand has accumulated alongside the western breakwater at Wanganui Harbour, but at nearby Kaitoke Beach 70 metres of recession is indicated by the ruins of a 1940 concrete gun emplacement out on the lower foreshore.

Tectonic uplift during the 1855 earthquake directly advanced coastlines in the Wellington district (Fig. 80), and hastened the progradation of the Hutt delta. At Hawke Bay the 1931 earthquake caused a seaward movement of high and low tide lines, coastal landslides, and resulted in the partial draining of Ahuriri Lagoon, near Napier (Marshall, 1933).

Coastline recession has predominated on the west coast of the Auckland Peninsula, except for the local development of spits and cuspate forelands near the entrance to Kaipara Harbour between 1852 and 1966 (Wright, 1969) (Fig. 81). On the North Kaipara Head, Gibb (1978) measured recession at 25.4 metres per year. Progradation at Manukau Harbour between 1840 and 1974 was mapped by Williams (1977). On the east coast of this peninsula, Schofield (1967) found cartographic evidence of progradation on the Mangatawhiri Spit between 1871 and 1934, supported by the stratigraphic evidence of a coke layer now buried beneath sand 60 metres inland from the present foredune and thought to have come from the debris of an 1879 shipwreck. The sand came from the adjacent sea floor. In 1957 this progradation came to an end, and the beach has since been cut back to behind the 1934 line. At Mount Manganui, Healy (1977) traced the progradation of Panepane

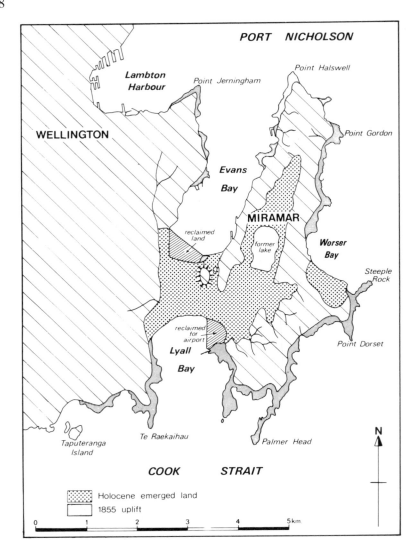

Fig. 80 The 1855 earthquake at Wellington, New Zealand, resulted in coastal progradation as shore areas were lifted above previous high tide level

Point, south-eastern Matakana Island, alongside Tauranga Harbour, by up to 500 metres between 1854 and 1954, followed by erosion, possibly related to port approach dredging and the arrival of larger waves during storms.

Recurrent volcanic eruptions, especially the Tarawera explosion of 1886, have delivered sandy material by way of rivers draining to the Bay of Plenty, the coastline of which has prograded as this sand, reworked by waves, was

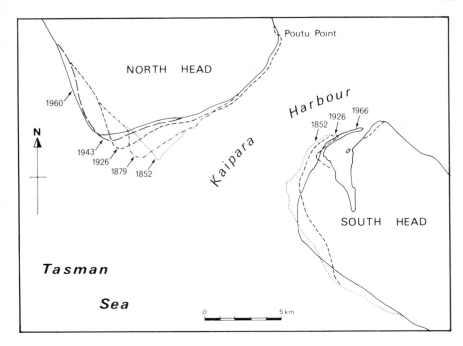

Fig. 81 Successive Admiralty Charts (1852, 1879, 1926) and air photographs (1943, 1960, 1966) have been used to trace the changes in sandy coastlines bordering the entrance to Kaipara Harbour, a large marine embayment on the west coast of North Island, New Zealand (Wright, 1969)

delivered to the shore. In Kawakawa Bay, near Gisborne, a sandy gravel beach fed with sediment from the Awatere River prograded 80 metres between 1899 and 1975. Gibb (1978) reported cliff recession averaging 2.25 metres per year in mudstones at Cape Turnagain, and 3.46 metres per year on conglomerates at Ngapotiki, near Cape Palliser, the latter being the most rapid cliff retreat measured in New Zealand.

In the South Island, progradation has been generally confined to river-mouth areas, for example, the gravelly deltas of the Westland coast and the shores of the Clarence delta in Marlborough. Receding cliffed coasts occur at numerous localities on both coasts, the recession being locally rapid in the volcanic deposits of the Taranaki coast. Sand and gravel drifting northwards along the west coast have accumulated south of Greymouth harbour breakwaters, but at Waimangaroa to the north Gibb (1978) showed how a sandy foreland with Holocene dune ridges has been truncated by retreat of its western shore between 1876 and 1976, and prolonged by spit growth to the north. Beach drifting continues northwards to Farewell Spit, which has been lengthened, and has shown accretion on the landward rather than the seaward side; Gibb (1978) reported growth of 68.9 metres per year on this spit. At Nelson the Tahunanui spit changed in outline between 1853 and 1970, a boulder bank to the north being breached at the present harbour entrance (McLean, 1978).

On the east coast of the South Island changes have been generally small. Accretion of northward-drifting beach material south of the Timaru Harbour breakwater between 1870 and 1970 was accompanied by retreat of beaches and cliffs to the north (Fig. 82). In Canterbury Bight, progradation near Lake Ellesmere has balanced erosion to the south-west, so that the coastline has been re-orientated (Armon, 1974), while on the Banks Peninsula sand washed up from the floor of Okains Bay prograded the beach by 230 metres between 1872 and 1969 (Fig. 83). Beaches on either side of the Kaikoura Peninsula prograded up to 40 metres between 1942 and 1974, the southern beach having received sand and gravel from the Kowhai and Kahutara Rivers, while that to the north has been supplied by the Hapuku River: on the other hand, the northern flank of the Hapuku delta showed erosion, and near Wairepo Lagoon on the Kaikoura Peninsula the beach was cut back 47.6 metres (Kirk, 1975). Retreat of cliffed coastlines at Kaikoura has been up to 0.24 metres per year.

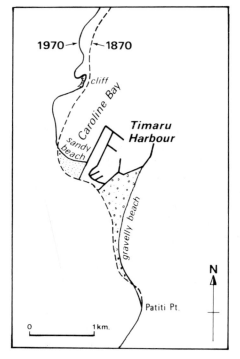

Fig. 82 The building of breakwaters to form Timaru Harbour has been followed by accretion of northward-drifting gravel and sand on the southern side, and continuing erosion of the coastline to the north, which has lost much of its former beach fringe (after McLean, 1978)

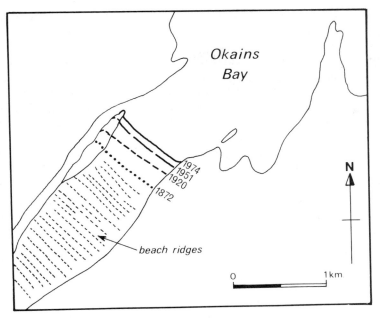

Fig. 83 Sand washed in from the sea floor has prograded the beach in Okains Bay, on Banks Peninsula, New Zealand (after McLean, 1978)

NEW CALEDONIA

In New Caledonia, coastline changes have generally been small, with only localized cliff recession and minor gains in mangrove areas, as on the Dumbéa delta (Baltzer, 1969), but substantial changes have occurred locally as the result of open-cast hill-top nickel and chromium mining during the past century, generating vast quantities of waste (mainly red clays, sands and gravelly deposits) which are moving down rivers to the coast (Bird, Dubois and Iltis, 1984). Mouths of rivers strongly affected by the arrival of this material include the Bouameu, Tinip, Ouaco, Népoui, Ouha, Poya, Tontouta, La Coulée and Rivière des Pirogues on the west coast, and the Kuébéni, Ouinné, Thio, Dothio, Nakéty, Canala, Ouango, Karoipa, Kouaoua, Poro and Néaoua on the east coast. On the Népoui delta, mangroves are advancing on to accreting tidal flats of red clay. The Thio delta has prograded in recent decades by the successive formation of bars, spits and sandy islands, and spits of sand and gravel have formed at Kouaoua. Changes unrelated to mining activities include the longshore growth of a shingle spit north-westwards from the delta of the gravel-yielding Koumac River, thereby advancing the coastline, and the growth and re-shaping of a shelly sand spit at the mouth of the Ouaième River in the north. Locally, sand and gravel washed onshore from fringing reefs have prograded coralline beaches, such as the Plage d'Oué near Bourail on the west coast.

There has been little change in recent decades on the coastlines of the Isle of Pines to the south, or on the emerged atolls of the Loyalty Islands to the north-east, of New Caledonia.

FIJI

The Fiji archipelago consists of two large mountainous islands, Viti Levu and Vanua Levu, some smaller high islands, and numerous low islands and coral reefs. In general volcanic formations predominate, but there are no active volcanoes. However, there have been occasional earthquakes and recurrent cyclones, and the generally low wave energy coastal environments have been subject to episodes of rapid change during storm surges.

The south coast of Viti Levu has alternations of promontories and mangrove-fringed bays, and is protected by reefs. The main changes in recent decades have been on the Navua delta coastline, where progradation has been interrupted by episodes of erosion and re-working of sediments by storm surges, and on the large swampy delta of the Rewa River, which is growing on the south-east coast. At Suva the coastline is now largely artificial, but offshore there are sand cays on the coral reefs: Makuluva is a cay that has migrated westwards in recent decades, erosional retreat of its eastern coast being balanced by progradation on the western shore.

On the east and north coasts of Viti Levu changes have been slow, mainly around mangrove-fringed deltas such as those of the Waimbula, Ndakuinuku, Ndawasamu and Mba Rivers. In the north-west there has been beach erosion at Sawena Bay, west of Lautoka, and the sandy coastline of Nandi Bay has retreated, but gains have exceeded losses on the Nandi delta in recent decades. There has been little change on the reef-fringed west coast, apart from some beach erosion at Momi.

Vanua Levu has similar steep coasts, with beach sectors and deltas growing at the mouths of the Ndreketi, Waileuvu, Lambasa-Nggawa, and Wainkoro Rivers. Beach erosion has been reported from a number of islands in the Fiji group, for example Kambara and Lakemba in the eastern islands, where McLean (1979) suggested that a diminishing supply of coralline sediment from reefs being widened by seaward growth of coral could be an explanation.

HAWAII

The Hawaiian Islands are volcanic in origin, and there are extensive steep coasts with receding cliffs, locally up to 900 metres high, where marine erosion is attacking structures built previously by deposition from volcanoes. Episodic progradation by lava flows has continued through the last few centuries on the island of Hawaii, where the south-east coast cliffs are cut into modern lava. During the past century there have been several episodes when lava flowed into the sea here, as in 1960, when the lava from Kilauea reached the coast near Cape Kumuhaki, adding 1.4 square kilometres of new terrain. Lavas from eruptions of Mauna Loa have repeatedly reached the sea south of Hookena, and nineteenth-century flows formed the promontories which border Kihola Bay (Shepard and Wanless, 1971). Beaches of black volcanic sand are found near the mouths of rivers that flow through lava plains, but the dominant beach material is calcareous, derived from marine organisms.

About 20% of the Hawaiian coastline is beach-fringed, and beaches are extensive (44%) on the island of Kauai, where a large cuspate foreland is prograding near Kekaha, in the south-west, while farther north Polihale Beach has widened by progradation. There are also widened beaches of calcareous sand, including material derived from coralline reefs, on parts of the west coast of Oahu, but the famous Waikiki Beach is now largely artificial, having been augmented by sand imported mainly from beaches on northern Oahu and western Molokai. On the north coast of Oahu, the beach at Waimea Bay has been cut back about 125 metres in the past century (Campbell and Hwang, 1982). On the west coast of Molokai is Papohaku Beach, one of those mined to provide sand for Waikiki and fine aggregate for the construction industry. This beach varies in extent from year to year, but there has apparently been sufficient newly-supplied biogenic sediment to maintain it despite the exploitation. From this it is inferred that beaches adjacent to richly-yielding biogenic environments may continue to prograde by deposition.

Detailed studies of Hawaiian beach systems by Moberley and Chamberlain (1964) indicated a total beach sand volume of 40 million cubic yards (for simplicity, their non-metric units are retained here), of which 14 million were located on Kauai and 10 million on Oahu. Seasonal gains and losses were up to several hundred cubic yards of sand per linear yard of beach per month. Subsequent monitoring (Campbell, 1972) showed that, over a decade, most beaches showed no significant gain or loss, but 14 beaches were identified as showing a net gain and another 14 a net loss over this period. Of the latter group, seven had a longer history of erosion. Chamberlain (1968) concluded that 'in the Hawaiian islands, beach sand is primarily acquired through the biological activities of marine organisms and primarily lost by paralic sedimentation'. Near reefs, biogenic sand production was of the order of 1000 to 5000 cubic yards per mile per year, while terrigenous sand supply was generally small, except near stream mouths. Occasionally, major short-term coastal changes, including rapid cliff recession and beach displacement, occur on the Hawaiian islands as a result of tsunami impact (Hwang, 1981).

TAHITI

On Tahiti there are only a few sectors of continuing beach and spit accretion, mainly near river mouths, on the generally steep, reef-fringed coasts, with cliffing only on promontories where the reef fringe is missing, as in the south-east. Mooréa, Raiatéa, Huahine, Bora Bora and Maupiti each shows similar features. No large-scale progradation has occurred during the past century.

OTHER PACIFIC ISLANDS

Coastlines of most of the other islands in the Pacific are either on volcanic formations, some (chiefly in the Solomon Islands, Samoa and the Galapagos), where active vulcanicity has changed coastal outlines, or coralline, usually uplifted reef structures. Beaches, whether of volcanic materials or reefal debris, are prograding only where there is a continuing sediment supply; many appear to be stable, but erosion has become a widespread problem, particularly evident where tourist resorts have been established (Kaplin, 1981). In 1972 a cyclone in the Ellice Islands produced a storm surge which added a new beach ridge of coralline sand and gravel 19 kilometres long and up to 4 metres high on the reef bordering the cay on the south-east of Funafuti Atoll. In subsequent years the new ridge has been driven shoreward up to 20 metres (Fig. 84), and will eventually be added to the existing cay coastline (Baines and McLean, 1976).

Steep cliffs have been cut into volcanic formations all round Easter Island, where recession can be measured with reference to prehistoric giant statues, one of which was undercut and destroyed between 1934 and 1955. The cliff of basic lava flows, 100 metres high, has here retreated about a metre in 10 years (Paskoff, 1978).

ATLANTIC OCEAN ISLANDS

In the Atlantic ocean, many of the islands are of volcanic origin, with cliffs cut back into lava and ash deposits, and local progradation of beaches where derived sands and gravels are accumulating: for example the growing spit at Porta Pim on Faial in the Azores. High receding cliffs in volcanic rock are seen on Ascension Island, St Helena, Gough Island and (with ice cliffs) Bouvet Island in the mid-Atlantic. Active vulcanicity has prograded coast-lines at Capelhino in the Azores and on Tristan da Cunha, where the 1961 earthquakes and eruptions were followed by rapid recession (5 metres per month in 1962) of volcanic cliffs, and the building of gravelly beaches and spits on the adjacent shores. Changes have been slow on the subantarctic rocky islands of the Scotia Group and the Falklands (Malvinas), except where stormy seas are cutting back low cliffs in lobes of periglacial solifluction deposits. Trindade and Fernando de Noronha are volcanic islands with

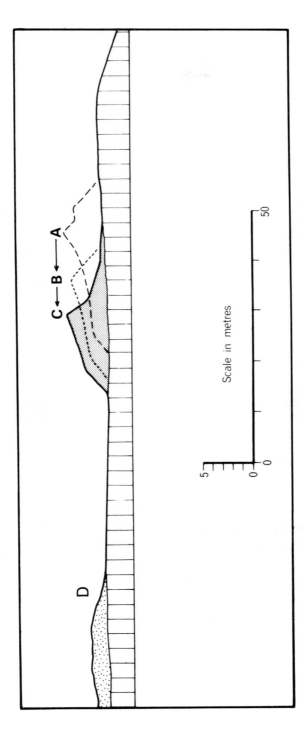

Fig. 84 Successive surveys of a ridge of coralline sand and gravel (A, B, C) thrown up on to the coral reef at Funafuti Atoll, South Pacific, in a 1972 hurricane showed that it has migrated landward, towards the pre-existing cay coastline (D) (after Baines and McLean, 1976)

Scale in metres

receding cliffs, but Bermuda consists of calcarenites, with minor cliffing and locally prograded sandy beaches receiving calcareous sand from the sea floor. The outlines of Sable Island off Nova Scotia have changed frequently, necessitating relocation of the lighthouse, but erosion has been compensated by deposition, the sandy islands maintaining their area of 30 square kilometres over the past two centuries, during which they have migrated up to 14.5 kilometres eastward (Cameron, 1965). This movement is partly due to the mobilisation of dunes following the reduction of their vegetation cover by introduced pony grazing (Owens and Bowen, 1977).

The Faeroes are rugged, mountainous basaltic islands with cliffed headlands and indented fiords: coastline changes have been very slow here, and on the hard rock islands of St Kilda and Rockall. Madeira has cliffs in volcanic formations and beaches of gravel and sand, and the same is true of the Selvagens Archipelago, the Canary Islands (where there has been some accretion at valley-mouth inlets), and the dry Cape Verde Islands. Fernando Po has steep forested coasts, with beaches and mangrove swamps in embayments, some of which have prograded in recent decades. Principe has cliffs retreating on its southern coastline, and São Tomé is another volcanic island, with retreating cliffs, and beaches of calcareous sand, some of which are gradually prograding.

INDIAN OCEAN ISLANDS

Coastline changes have been generally minor on high volcanic or granitic islands in the Indian Ocean such as the Comoros Archipelago, Réunion, Mauritius and Rodriguez. There are some retreating cliffs and a few sectors where beach accretion or mangrove advance is prograding the coastline, but extensive fringing reefs prevent strong wave action. Occasional cyclones have caused rapid recession of coastlines, as in Mauritius in 1960 (McIntire and Walker, 1964), when beaches were scoured away or driven landward. In northern Seychelles the granitic islands of the Mahé group have steep, rocky shores, and intervening sandy beaches, but changes have been very slow. In the Maldives the islanders have endeavoured to enlarge their small coral cays by building groynes to trap sand drifting in the nearshore shallows.

Uplifted coral islands, such as Christmas Island, Aldabra, Cosmoledo, Astove, and Assumption are each bordered by slowly receding cliffs cut in coral limestone. Low sandy cays on reefs in the Maldives, the Laccadives, the Amirantes, in the Moçambique Channel, on Chagos Bank, and at Cocos are all subject to coastline changes, erosion of one sector usually being balanced by accretion on others. In many cases erosion has been slowed down by the presence of beach rock.

ANTARCTICA

Ice coasts are widespread in the Antarctic, but information concerning their advance or recession is sparse. Research on the form and dynamics of the ice sheet may yield data on changes at coastal margins: it appears that the decline of Antarctic ice volume in the 1970s has subsequently been reversed (Kukla and Gavin, 1983). On sectors where rocky shores are exposed in summer there are cliffs and beaches, some of the latter being directly nourished by angular gravels generated by periglacial solifluction, or from melting glaciers. These features are similar to those seen in the Arctic, but the deltaic coasts of northern Canada and Siberia are not matched in the Antarctic. On subantarctic Heard Island, active vulcanicity has generated abundant sand and gravel, carried down to the coast in glaciers and meltwater, and delivered to beaches which drift eastwards to accumulate in a large spit trailing from the eastern end. Much of the rest of the island is cliffed, particularly on the western coast.

CHAPTER THREE

Categories of coastal change

The coastline changes described in the preceding chapter fall into a number
of categories, which will now be reviewed. It is evident that the proportion
of the world's coastline that has retreated during the past century exceeds
that which has advanced, although extensive sectors have remained stable,
or have shown no definite evidence of advance or retreat. While a few
coastlines have advanced or retreated by more than 100 metres per year, on
the world scale a gain or loss of more than 10 metres per year has been
exceptionally rapid, and very few coastlines have changed by more than ±
1 metre per year.

In reviewing the various categories of coastline change, some reference
will be made to possible geomorphological explanations. Further research
will elucidate these, identifying and ranking the various factors and processes
that have led to erosion or deposition on particular sectors of coastline.
Detailed investigation of a particular changing coastline usually indicates that
several factors must be taken into account: a single-factor explanation usually
turns out to be an over-simplification. The examples given in this review,
together with the references, should be useful to students working on a
particular sector of coastline where changes have taken place.

CLIFFED COASTLINES

The rates of recession of cliffed coastlines have varied with their elevation,
rock resistance, geological structure, and incident wave energy. Cliffs in
hard, massive rocks have changed very little during the past century, even
where they are exposed to stormy seas, while cliffs in soft materials have
been cut back very rapidly. The most rapid changes have occurred where

volcanic activity has deposited ash (pyroclastic sand and gravel) in a high wave energy environment. This has occurred, for example, in the Aleutian Islands, on Hawaii, at Surtsey off Iceland (Fig. 15), on Krakatau (Fig. 64), and on Isla San Benedicto, off the Pacific coast of Mexico, where Richards (1960) reported initial cliff recession rates of up to a metre a day. Rapid recession has also occurred on cliffs cut in glacial drift subject to seasonal wave attack on Arctic coasts, notably in Siberia, where Zenkovich (1967) quoted rates of cliff retreat of up to 100 metres per year on tundra bluffs weakened by the summer thaw. Here, as elsewhere, rates of cliff recession have varied from year to year, with episodes of rapid erosion during stormy periods. Thus on the glacial drift cliffs on the Polish coast, Zenkovich (1967) calculated an average recession of a metre a year, but up to 5 metres have been lost during a single storm. The building of sea walls to halt cliff recession has been extensive on such coasts, and where these have been successful (as on Heligoland) the coastline is now stable.

Variations in rates of cliff retreat can result from changes in the widths of the beach, as in Holderness, where erosion slackens as lobes of beach material drift by, and revives with the passage of intervening lower 'ords'. As a cliff recedes, relatively resistant shore outcrops may be exposed, impeding wave attack and slowing down the erosion rate. The progress of cliff recession may also be related to water depths in the nearshore zone. On the Holderness coast, and in Western Jutland, coastline retreat has probably been accompanied by nearshore sea floor degradation, so that the transverse profile is maintained as it migrates landward. Subaerial processes, notably runoff after heavy rain, have contributed to the dissection and cutting back of many cliff faces in varying degrees, but cliff recesssion is maintained by basal marine erosion.

GLACIATED AND PERIGLACIATED COASTLINES

Recession of ice coasts as glaciers melt has been well documented on the Gulf of Alaska, particularly in Icy Bay, where the glacier snouts retreated at up to a kilometre per year between 1904 and 1971 (Fig. 8). On Ellesmere Island, in northern Canada, on the Greenland coast, and around Antarctica the seaward margins of glaciers and ice shelves have alternately advanced and retreated during the past century, but little information is available on the extent and duration of such alternations. In southern Chile an ice coast has advanced, then retreated, at the Lagune de San Rafael within the past 200 years.

There has been deposition of glacifluvial sediment on prograding deltaic plains in front of melting glaciers in recent decades in several bays and inlets on the Gulf of Alaska, notably at Juneau. Rivers augmented by melting ice and glacifluvial outwash have built deltas or supplied sand and gravel to prograde beaches near their mouths, as on the Westland coast in the South Island of New Zealand. Where the melting of ice has been hastened by volcanic eruptions, as in south-east Iceland, and on Heard Island in the

Southern Ocean, large quantities of torrential outwash have been delivered to the coast, and subsequently incorporated in beaches, spits, and barrier islands. Periglacial activity has supplied gravelly solifluction debris to beaches on the coasts of Arctic Canada, notably on Devon and Somerset Islands, on the shores of Davis Strait in Greenland, and on subantarctic islands such as the South Orkneys, South Shetlands and South Georgia.

EMERGING AND SUBMERGING COASTLINES

Changes in the relative levels of land and sea have led to gains and losses on coastlines during the past century (Bird and Paskoff, 1979). Isostatic uplift of formerly glaciated areas in northern Canada and Scandinavia has resulted in the emergence of former nearshore areas, as in the Gulf of Bothnia, where the Swedish and Finnish coastlines have locally advanced more than 100 metres in this period (Figs. 16, 18). Similar features are seen around the Caspian Sea, as the result of its lowering between 1929 and 1977 (Fig. 49): outlying islands have expanded, and some shoals have emerged as islands (Fig. 50). Sudden emergence during earthquakes has advanced coastlines by several metres on Montague Island and the Copper River delta in Alaska (1964), Talcahuano in Chile (1855), Bartin and the Orontes delta in Turkey (1958), Pasni on the Makran coast in Pakistan (1945), Cheduba Island in Burma (1762), Tokyo Bay in Japan (1923), and Wellington (1855) (Fig. 80) and Napier (1931) in New Zealand. Such emergence has often been followed by shoreward drifting of sea floor sediments through shallowed water to prograde beaches, as at Kalajoki in Finland and on the Makran coast in Pakistan.

By contrast, submergence resulting in coastline retreat as the sea invades the land margin has been widespread, but generally minor. Many tide gauge records show a rise of mean sea level of several centimetres during recent decades, but the effects of this slight rise have been recognized only locally, as a possible contribution to beach erosion and the recession of seaward margins of many salt marshes and some mangrove swamps (see below). It is only when accompanying land subsidence has accelerated this modern marine transgression, as in Louisiana, where abandoned Mississippi delta lobes show retreating coastlines, eroding as they are submerged, that the changes have been marked. In the Netherlands, subsidence augmenting sea level rise would have set back the coastline were it not for the success of Dutch reclamation works. Subsidence of the Venice area, due partly to subsurface compaction following the pumping-out of groundwater, has accelerated beach retreat on the Lidi, and led to the cutting-back and dissection of submerging salt marshes in the Venetian lagoon. Beach retreat and swamp submergence on the north coast of the Sepik delta, New Guinea, are probably the outcome of local subsidence accompanying a sea level rise.

Sudden subsidence due to earthquakes has led to coastline recession at Anchorage, Alaska (1964), in Southern Colombia (1979), at Concepción,

Chile (1960), and on the Indus delta (1819, 1845), while in New Guinea coastal swamps were flooded to form Sissano Lagoon as the result of the 1907 earthquake. Landslides triggered by the 1964 earthquake formed temporary lobes on the coast at Turnagain Heights, Alaska, and recurrent tremors have produced similar features in the southern Philippines and along the north coast of Irian Jaya. Submarine mass movements caused sudden recession of the delta fronting the Valdez glacier in Alaska, and collapse of the wall of a submarine canyon led to rapid beach retreat east of Abidjan, on the Ivory Coast.

In the Great Lakes, oscillations of water level during the past century have resulted in short-term advance and retreat of the coastline, falling levels promoting shoreward drifting of sediment and beach accretion, rising levels leading to marginal submergence and beach erosion, and stimulating cliff retreat. However, the relationship between sea level changes and the advance and retreat of coastlines is complicated by other factors, especially movements of sediment onshore, offshore or alongshore. In Denmark, erosion is prevalent along the emerging coast of north Jutland, and progradation is continuing, despite submergence, on the barrier islands of south-west Jutland (Fig. 23).

VOLCANIC COASTLINES

Vulcanicity has resulted in coastline progradation both directly, where lava and ash have been deposited on the shore, and indirectly, by feeding large quantities of sediment into rivers, thereby accelerating delta growth or the progradation of fluvially nourished beaches. Successive eruptions have produced the lavas that advanced parts of the south-east coast of Hawaii, while vulcanogenic deposition has prograded coastlines in Guatemala (Rio Samala delta), south-east Iceland, the Philippines (South-east Luzon, North Mindanao), Papua New Guinea (Mount Lamington) and the Bay of Plenty, New Zealand (after the Tarawere explosion in 1886). Many oceanic islands are entirely of volcanic origin; some, like Surtsey, have appeared only recently (1963–7), others, like Bogoslof in the Aleutians, have been eroded away completely in recent decades. Santorini in the Mediterranean, and Krakatau in Indonesia, are islands that were modified by explosive eruptions, leaving caldera-side cliffs and mantles of rapidly eroding ash: in both cases a newer volcanic island has appeared and is developing within the caldera (Fig. 65).

LANDSLIDES

Landslide lobes and rock falls temporarily prograde steep and cliffed coastlines, and are subsequently cut back by marine erosion (Fig. 29). Examples of this have been recorded at Tillamook Head, Washington; Pacific Palisades, California; on the Norfolk and Dorset coasts in England; at Rosnaes and Helgenaes in Denmark, and on the Bulgarian and Crimean coasts of the

Black Sea. Some lobes have persisted long enough to act as natural break-waters against which drifting beach sediment accumulates. On the Dorset coast, arcuate boulder ridges formed at the margins of landslide lobes persist in the nearshore area after the softer material has been washed away by wave action, commemorating the seaward limits of the landslide.

DELTAIC COASTLINES

Rapid accretion has occurred on prograding deltas, with rates of up to 80 metres a year at the South-west Pass of the Mississippi delta and on the new Brazos delta in Texas, and up to 70 metres a year on the Irrawaddy delta. Vestappen (1966) quoted evidence of progradation by up to 180 metres a year on tropical deltas in Indonesia. The progradation of deltaic land (i.e. above high tide level) depends on deposition from river floodwaters or the formation of wave-built beach ridges and spits, sometimes capped with wind-blown sand, along the seaward margins. Some deltaic shores advance by the seaward spread of sediment-trapping salt marshes (e.g. Fraser delta, British Columbia) or mangroves (e.g. parts of most tropical deltas), the formation of supratidal land then depending on accretion from floodwaters. As well as depositing sediment at river mouths, floods supply sediment that can be distributed alongshore to prograding beaches: the Columbia River on the Pacific coast of the United States, the Rio Grande de Santiago in western Mexico, the Danube in the Black Sea, and the gravelly rivers of the west coast of South Island, New Zealand, each exemplify this.

When fluvial sediment supply is diverted, the pattern of coastal accretion changes. There are several examples of changes in the location of a river outlet during floods, the major example being the Huanghe, China (1852), while others are the Rio Sinu in Colombia (1942), the Ceyhan in Turkey (1935), the Medjerda in Tunisia (1973) and the Cimanuk in Java (1947). In each case the abandoned delta has been cut back by erosion, while the new outlet has become the site of deltaic progradation (Figs. 53, 67). The same effect has obtained when the diversion has been artificial, by way of a canal cut to a new outlet, as on the Brazos delta in Texas (1938), the Rioni in the eastern Black Sea (1939), the Shinano in Japan (1922) and the Cidurian in Java (1927). A new delta has developed on a previously retreating coastline since the cutting of a canal from the Mississippi to artificially revive the Atchafalaya distributary.

Rates of progradation of deltas and coastal plains have accelerated when fluvial sediment yields have increased as the result of soil erosion due to deforestation, over-grazing, or cultivation of steep hinterlands. This has been widespread, especially in the wet tropics: in Java, for example, soil erosion has led to increased rates of accretion during the past century in and around Jakarta Bay, on the north coast deltas, and in the Segara Anakan embayment, which is shallowing and shrinking as the result of sedimentation, mainly from the Citanduy River (Fig. 67). Accelerated deposition at and around river mouths has also been reported from several Caribbean islands, from Brazil, Chesapeake Bay in the United States, the Gulf of Cambay in India, parts of Malaysia, the Philippines, Japan and New Zealand. In some areas, increased fluvial sediment yields are due to hinterland mining activities, especially hydraulic sluicing, which augmented river loads in the Sierra Nevada a century ago and has led to rapid accretion in San Francisco Bay, and open-cast quarrying, which has generated mining waste spilling into rivers and thence to estuaries, deltas and prograding beaches: for example at Par and Pentewan in Cornwall, at Chānaral in Chile, on the Tenasserim coast in Burma, at Jaba on Bougainville, and extensively around New Caledonia, especially on the Népoui and Thio deltas.

There have been many examples of the onset of erosion on deltaic coastlines following dam construction on rivers and consequent reductions in water flow and sediment yield to river mouths. The best known example is the Nile delta after completion of the Aswan High Dam in 1964 (Fig. 54), but a similar sequence has been seen on the coastlines of the Rhône, the Dnieper, the Dniester, the Volga, the Volta and the Zambezi: most of the world's major rivers now have dams, and the effects of reduced sediment yield will become more widespread in the next few decades. Around the Mediterranean, particularly in Italy and Greece, erosion of deltas and depletion of beaches formerly supplied with sediment from rivers in recent decades has also been attributed to the effects of dredging sand and gravel from river channels and to diminished runoff due to irrigation schemes, soil conservation works, and afforestation. In parts of Greece and Turkey the diminution of fluvial sediment yield is thought to be due partly to the increasing exposure of hard rock in areas where the soil cover and weathered mantle have been eroded completely.

BEACHES, SPITS AND BARRIER COASTLINES

Changes on beach-fringed coastlines result from variations in the rate of supply of sediment from various sources, or in the pattern of sediment losses from a beach system (Fig. 85). These variations may be due to several factors, including changes in sea level relative to the land, climatic variations, geomorphological changes, or modification by human activities.

Progradation of beaches and growth of spits supplied with sediment eroded from nearby cliffs (by wave attack or surface runoff), and transported alongshore is well illustrated in areas dominated by glacial drift deposits, as in Puget Sound and the Strait of Georgia on the west coast of North America, New England and the Canadian Maritime Provinces on the east coast of North America, the shores of the North Sea, Denmark, and the Southern Baltic. Sand Point, in Tomales Bay, California, has grown by accretion of sand derived from cliffs to the north, and beaches in bays on the central coast of Albania have prograded by receipt of sediment eroded from flanking cliffed promontories.

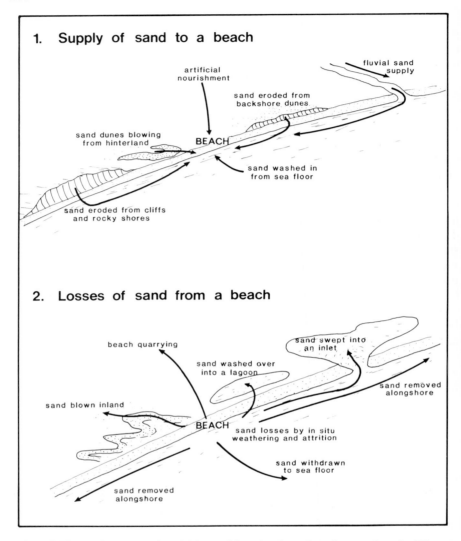

Fig. 85 The various ways in which sand is gained, or lost from, a beach. When gains have exceeded losses beach progradation is likely to occur, while losses exceeding gains lead to a lowering, flattening, and cutting back of a beach

When sediment supply from eroding cliffs diminishes, adjacent beaches may become depleted. During a prolonged stillstand of sea level, cliff recession will declerate if a shore platform or nearshore shelf develops and widens, diminishing incident wave energy, and thereby reducing the sediment supply to adjacent shores. Zenkovich (1976) reported evidence of this from the Soviet Black Sea coast. Where cliff recession has been halted by the construction of sea walls, as at Brighton and Bournemouth in southern England, on the Melbourne coast in Australia and at Byobugaura in Japan, the sediment supply to adjacent beaches has come to an end, and beach erosion has ensued. Cliff protection was a factor in the reduction of longshore sediment supply to Ediz Hook, Washington, and the consequent onset of erosion there.

Man's impact on beaches may be either constructive or destructive. The spilling of waste from coastal quarries has provided material to prograde beaches at such locations as Hoed in Denmark, Porthoustock on the coast of Cornwall, and Rapid Bay in South Australia. At Workington in Cumbria the dumping of coal-mine wastes and slag from a steelworks has locally augmented beaches. By contrast, the quarrying of sand or gravel from beaches (Fig. 86) not only depletes them, but also may result in accelerated erosion of cliffs and coastal terrain formerly protected from strong wave action by the beach fringe, as at West Bay on the Dorset coast.

Examples of beaches prograding where winds are driving sand from the hinterland into the sea are found on San Miguel Island, off California, in the southern part of the Bahia de Paracas, Peru, and at Castillos in Uruguay. Desert sand from the Sahara is nourishing beaches from southern Morocco to Mauritania, while barchans from the Namibian desert are reaching the coast of Tiger Bay, Angola (Fig. 56), and south from Lüderitz, and Arabian

Fig. 86 Sand and gravel extraction from beaches has contributed to coastline erosion in many parts of the world. This photograph, taken near Klim, on the north coast of Jutland, Denmark, records the situation the day before such extraction became illegal on this coastline. Photo: Eric Bird (May 1974)

desert sands are blowing on to beaches in south-eastern Qatar. On the Cape Coast in South Africa, sand dunes spilling eastwards across headlands, as at Britannia Point, maintain and prograde lee-side beaches, and similar features are seen on south-facing coasts in Australia, as at Woolamai and Yanakie in Victoria, where dunes are spilling across isthmuses. Beach accretion comes to an end in such a situation where the hinterland sand supply runs out, or if conservation measures, such as marram grass planting, stabilize the spilling dunes. Beach erosion at Port Elizabeth is partly related to the re-vegetation of the dunes that formerly drifted over Cape Recife, while the success of marram grass planting on the Woolamai dunes in Victoria in 1970 has been followed by the onset of beach erosion in Cleeland Bight (Fig. 75).

Beach progradation in the lee of nearshore shoals has been reported from Lacosta Island, Florida, the coast of South Virginia, and County Wexford, Eire. Migration of prograded sectors as such shoals move alongshore has occurred at Benacre Ness, East Anglia, and False Cape, Virginia, and along the swampy coastlines of French Guiana and Surinam. In each case there has been erosion of the formerly prograded sector after the protecting shoal has moved on.

Examples of beach progradation following a decline in the vegetation which formerly held sediment in place on the adjacent sea floor have been documented on the Danish island of Kyholm and on the Mediterranean coast in Provence.

Variations in rates of beach progradation and alternations of erosion and deposition may be correlated with short-term fluctuations in fluvial sediment yield related to rainfall incidence. In southern California the beach erosion prevalent in the dry years 1939 to 1968 has been partly offset by rapid sedimentation from river floods in wet years 1969, 1978, 1980, and 1983 (Orme, 1985). In Chile the lobate delta built by the Arica River during a flood in 1973 has been cut back by marine erosion in ensuing dry years. At Caraguatatuba, Brazil, a 1968 flood greatly augmented the sand supply to the coast, prograding beaches that had previously been eroding (Fig. 5).

Coastline recession as the result of storm surges has been documented around the North Sea and in the Bay of Bengal, while hurricanes or cyclones have generated waves that cut back cliffs and beaches and eroded swamp-lands along the Gulf and Atlantic coasts of the United States, in Hong Kong, and on Mauritius. About half the beach recession at Miami between 1884 and 1944 occurred during the 1926 hurricane.

Longshore movement of sediment results in various kinds of coastline change. Longshore spit growth, as at Orford Ness, or the Langue de Barbarie in Senegal, effectively advances the coastline, as does accretion alongside headlands (Point Dume, California; Apam, Ghana; Cape Wom, Papua New Guinea) or artificial breakwaters. There are numerous examples of beach accretion updrift of harbour breakwaters accompanied by erosion downdrift: Lagos in Nigeria (Fig. 55) and Durban in South Africa are well-known examples. Accretion on both sides of such breakwaters occurs where longshore drifting converges or alternates, or where sand has moved in from the sea floor on either side of an ebb jet: examples include Rogue River in Oregon; Newport, California; the Swina Inlet mouth in Poland; Umuiden, Netherlands; and Lakes Entrance in Victoria, Australia (Fig. 76). Beach accretion has occurred behind wrecked ships at Sukhumi, on the Soviet Black Sea coast. Natural beach progradation has attached the Chilean island of Quintero to the mainland, and accretion beside a breakwater has formed a tombolo at Morro Bay, California. Repeated breaching and re-forming of sandy tombolos has been charted at Gabo Island and Broulee Island, in south-eastern Australia.

Movement of beach material alongshore from an eroding to an accreting sector can result in straightening of the coastline, as at Giulianova in Italy, Kujukurihama Bay, Japan (Fig. 61), and Seven Mile Beach, South-east Tasmania, all gently-curving embayments where in recent decades central sector progradation has balanced erosion at either end. In the Andalusian Bight (south-western Spain) the reverse is the case: the central sector has been cut back while the sandy coastlines north of Mazagon and south of Matalascañas have prograded. Among examples of the re-orientation of a beach within an embayment is the curved northern end of Canterbury Bight, New Zealand, which has been re-shaped by erosion at Lake Ellesmere and accretion to the north.

Erosion has been strong on coastal salients from which longshore drifting diverges, as at Kolobrzeg, Poland; north-east East Anglia; Budaki in the north-west Black Sea; and on the Niger delta.

Longshore spit growth on successively set-back alignments related to adjacent cliff recession has occurred at Cape Cod, at Cape Henlopen along-side Delaware Bay (Fig. 13), on the Hel spit in Poland, Hurst Castle spit in southern England, and the Pointe de la Coubre in western France. On the Black Sea coast near Odessa, recession of the Tendrovskaya Kosa coastline has been accompanied by spit growth at both the eastern and western ends. Cuspate forelands eroding on one flank and accreting on the other migrate alongshore: the major capes of the Atlantic coast of the United States have shown this trend, as have the Darss Foreland in East Germany, Grenen on Jutland, Flakket on Anholt, Winterton Ness, Benacre Ness and Dungeness in England, the Magilligan foreland in Northern Ireland, and Cape Pitsunda on the Soviet Black Sea coast. Another example is Point Hope, Alaska, but Point Barrow to the north now shows erosion on both flanks (Fig. 6).

Spit prolongation can be very rapid: up 125 metres a year on Godavari Point, in India. Addition of recurves to Pointe de la Coubre in France advanced the coastline southward by 36 metres a year between 1881 and 1948. Progradation on Clatsop Spit in Washington, USA, averaged 20 metres per year, and there are several examples of gains (and matching losses) of more than 10 metres per year on the ends of spits and the flanks of cuspate forelands.

In the course of its work, the IGU Commission on the Coastal Environment gave particular attention to the problem of erosion on the world's sandy coastlines, and an analysis of the results is given in the following section.

EROSION OF SANDY COASTLINES

On the basis of information supplied to the Commission on the Coastal Environment, about 20% of the world's coast is sandy and backed by beach ridges, dunes, or other sandy depositional terrain. Of this, more than 70% has shown net erosion over the past few decades, and less than 10% sustained progradation, the remaining 20–30% having been stable, or having shown no measurable change. More detailed studies of sandy coastlines have increased the allocation to the receded category. The past century has thus been a phase of widespread erosion, with many formerly prograding coastlines, notably on sandy barriers and deltas, showing net retreat (Fig. 87). Apart from sectors where the coast has advanced because of land reclamation or sediment accumulation alongside artificial structures, the main sectors of progradation have been on beaches supplied with sediment from river mouths, on growing parts of spits and cuspate forelands, and on emerging coasts, notably in Northern Canada and Scandinavia, and in the Caspian Sea. Away from river mouths, eroding cliffs and spilling dunes, sandy coastline progradation has been restricted to coasts adjacent to sea floor sand sources. These include sand shoals off south-west Jutland, the south-east coasts of the USA, and in the Rio de la Plata west of Montevideo; glacial drift off eastern Scotland and around the Baltic; submerged deltas off south-west Africa; the sandur off south-east Iceland; and biogenic sediments such as oolites from the Bahama Banks, shells from the Sea of Azov, the Caspian Sea, and the southern Arabian Gulf, and coralline sand and gravel from

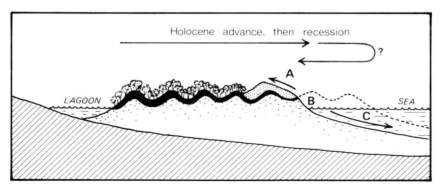

Fig. 87 Many sandy barrier coastlines that prograded in Holocene times, with successive formation of beach or dune ridges, now show evidence of recession on their seaward margins. Some of the sand has been washed or blown onshore (A), some has moved away alongshore (from B), and some has been withdrawn to the sea floor (C)

reefs around Hawaii and many other islands in the Indian and Pacific Oceans. The extensive sandy barrier coasts of Brazil, the Gulf of Mexico, the Atlantic coasts of the United States and Canada, the southern Baltic, West and South Africa, India and Sri Lanka, the Soviet Union, Northern Japan, and Australia have all shown recession during the past century. In some cases the barriers are transgressive features, migrating landward as the result of dune drifting and overwash by storm surges (e.g. Atlantic USA from the Carolinas to Rhode Island); in others there is evidence of a transition from earlier Holocene progradation to modern coastline retreat during the past century, as on the Ninety Mile Beach in south-east Australia, where beach and dune ridges have been removed by erosion. As blowouts develop and coalesce behind such shores, and washovers occur on narrow sections, these formerly prograded barriers may evolve into transgressive formations.

This modern prevalence of erosion on the world's sandy coastlines has been demonstrated and analysed in a number of papers (e.g. Bird, 1979, 1981a, b, 1983; Bird and Paskoff, 1979; Paskoff, 1983). Several hypotheses have been put forward. It is widely held that a world-wide rise of sea level has taken place during the past few decades, at an average rate of just over a millimetre a year (Fairbridge, 1966). This may be part of a new oscillation of ocean levels of the kind that has occurred frequently through Quaternary times; or it may be correlated with a global increase in atmospheric and oceanic temperatures, which in turn may follow increased carbon dioxide in the atmosphere as a consequence of large-scale combustion of fossil fuels in the modern industrial era. Bruun (1962) has argued that on an 'equilibrium coast' a sea level rise would result in a landward migration of the transverse shore profile, with consequent coastline retreat, and the transference of sand from the beach to the nearshore zone (Fig. 88). This hypothesis fits the erosional phases around the Great Lakes, which correlate with periods of

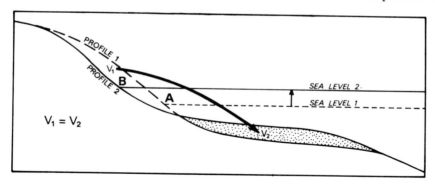

Fig. 88 The response of a beach which has attained equilibrium with nearshore processes at Sea Level 1 to a rise in level is coastline retreat (A to B) as sediment is removed from the beach face to the nearshore area (V_1 to V_2), thereby restoring the transverse profile (Dubois, 1980, modified after Bruun, 1962: see also Schwartz, 1967)

rising lake level, but it is doubtful whether so small a change in the level of the oceans is sufficient to account for predominance of beach erosion, although it certainly would have been a contributory factor. Sandy coastlines are in retreat through a variety of coastal environments, ranging from the low to moderate wave energy Baltic coasts to the high wave energy ocean shores of the southern continents; within Australia, for example, they range from the macrotidal north-west coast to the microtidal south-east coast, where measurements of mean sea level fluctuations from year to year exceed the dimensions of the long-term world-wide rise. Moreover, the erosion of sandy

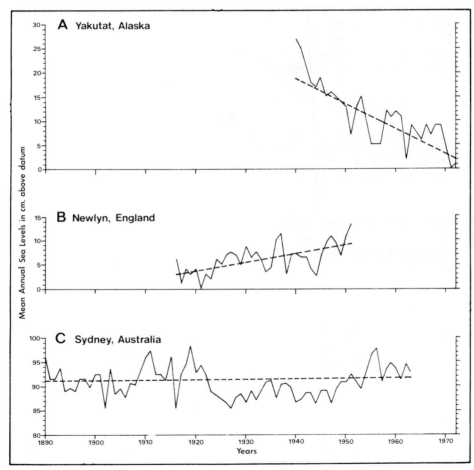

Fig. 89 Mean annual sea levels determined from tide gauge records show trends (pecked lines) of coastal emergence at Yakutat, Alaska, where the land is rising because of postglacial isostatic recovery, coastal submergence at Newlyn, south-west England, where the land may be subsiding to augment a sea level rise, and to no nett change at Sydney, Australia

coastlines began at different times in different places: it cannot be correlated simply with a sea level rise. Many coasts do not show evidence of a contemporary sea level rise, perhaps because of land uplift (Fig. 89). Yet there has been erosion on some sandy coastlines where emergence is in progress, as on the Indalsälven delta in northern Sweden, an area of isostatic land uplift, and on parts of the coastline of the Caspian Sea as levels fell between 1930 and 1975.

There is no doubt that artificial structures, such as sea walls, groynes and breakwaters, have led to beach erosion locally (as in Half Moon Bay, California, Newhaven in England (Fig. 90), and at Dutton Way, near Portland Harbour in Victoria, Australia) or that dredging of port approaches (as at Tauranga Harbour, New Zealand) and extraction of minerals, or sand and gravel, from nearshore areas can augment wave attack and thus deplete adjacent beaches. Extraction of beach sands has been followed by rapid erosion on many coasts, especially in the West Indies, around the Mediterranean, and in Europe (Fig. 86), but the extent of erosion on the world's sandy coastlines has been too great to be explicable simply in terms of human interference. Erosion is in progress on sparsely populated and little-developed sandy coasts in Brazil, South Africa, and Australia, as well as along the much-modified coast of the eastern United States. Although the building of dams and impounding of reservoirs on river systems has undoubt-

Fig. 90 The building of a breakwater to protect the entrance to Newhaven Harbour, southern England, in the nineteenth century has been followed by accretion of eastward-drifting beach shingle, 'and the degradation of the clay-capped chalk cliff, where the widened beach prevents it being attacked by wave action. Photo: Eric Bird (March 1962)

172

edly diminished fluvial sediment yields and has led to the onset of erosion on deltas, and along fluvially nourished beaches, erosion has been at least as marked on the barriers of eastern United States and south-eastern Australia, which were not produced by fluvial sand supply.

Reduction in the supply of sand from eroding cliffs, either because of a diminution in their erosion rate or because of the artificial stabilization of cliffed coasts, may have contributed to beach erosion locally, but cannot explain the onset of erosion on beaches far from cliffed sectors. The same is true of any reduction in the supply of wind-blown sand from hinterlands, which, as has been indicated, is a local explanation of beach erosion, for example at Port Elizabeth in South Africa. Losses of sand landward from backshore dunes may accelerate, but cannot initiate, sandy coastline recession (Fig. 91). A contributory factor to beach reduction is the gradual diminution of volume that occurs on beaches no longer receiving a sediment supply, because of secular weathering, solution or attrition of the component sand particles: as they diminish in calibre the beach profile becomes lower and flatter, and finer grains are more readily winnowed by waves, currents or wind action, so that eventually erosion prevails and the coastline retreats (Fig. 92).

Another suggestion has been that beach erosion is due to an increase in storminess—in the frequency and/or severity of storm wave activity in coastal waters—resulting in erosion of sandy coastlines that would otherwise have

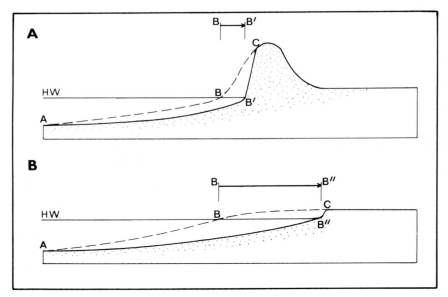

Fig. 91 On coasts where a foredune stands behind a beach, erosion of a specific volume of beach material results in less recession (B to B′) than where there is no foredune (B to B″), or where a foredune has drifted inland, or been removed by quarrying

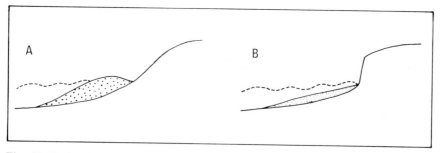

Fig. 92 Reduction of beach volume by attrition (the diminution of particle size, e.g. from pebbles to sand) results in lowering of a beach, so that wave action is able to penetrate to the backshore and attack the coast, initiating or accelerating cliff recession. A similar effect may be achieved where beach volume is reduced by the leaching of soluble constituents (e.g. carbonates) from the beach material

remained stable, or continued their earlier progradation. While there is some evidence for historical variations in storminess, for example around the British Isles (Manley, 1952), a trend towards increased storminess would have to be almost world-wide to account for erosion on geographically dispersed sandy coastlines. Climatic systems that permit increased storminess on one sector generally result in less stormy conditions elsewhere, so that even within a small region such as south-eastern Australia (Thom, 1978), this factor, though possibly contributory, is only one of several relevant to the onset of erosion. Bryant (1983b) correlated advance and recession of the beach at Stanwell Park, New South Wales, with fluctuations in mean sea level recorded at Sydney, which in turn reflect variations in atmospheric pressure patterns over the Southern Hemisphere: he correlated beach recession with rising sea levels (and associated rising beach water-table levels) rather than with storminess, and beach accretion with episodes of falling sea level.

Russell (1967) put forward the hypothesis that beach and barrier progradation was the outcome of shoreward sand movement, initiated during the Holocene marine transgression, which brought the sea to approximately its present level some 6000 years ago, and continued after the modern stillstand was established. The sea floor sand supply then consisted of material deposited by rivers discharging to low late Pleistocene sea levels, the relics of beaches and barriers left behind in the preceding sea level fall, and possibly aeolian deposits. During the oscillating Holocene marine transgression, part of this material was collected and carried shoreward, and when the sea level became relatively stable this was built up into prograding beaches and barriers. Sectors of such coasts not receiving fluvially supplied sand would eventually attain a transverse profile that must migrate landward to compensate any losses from the shore system, whether by wind action to hinterland dunes, longshore drift, movement into estuarine or lagoonal 'traps', or withdrawal to deep water offshore. This would explain why the

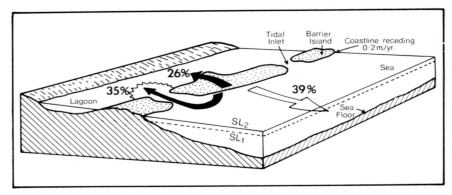

Fig. 93 The sandy coastline of Rhode Island, USA,> has retreated 0.7 metres per year between 1939 and 1975. Fisher (1980) has calculated that 35% of the sand volume lost has been washed into tidal inlets, 26% has been washed or blown over barrier islands into lagoons or swamps, and the remaining 39% has been submerged, and removed offshore to the sea floor

erosion was initiated at different times on the world's sandy coastlines, according to the quantities of sea floor sand available, the extent of natural replenishment (if any), and the rate at which shoreward drifting had taken place.

This hypothesis fits many of the features shown by sandy barrier coasts, but no one hypothesis can account for the prevalence of beach erosion in the variety of environments around the world's coastlines. Several factors have contributed to the predominance of erosion on such coasts during the past century, and it has not been possible to correlate the modern prevalence of beach erosion simply with sea level changes, or climatic or tidal cycles, although each of these has been contributory. The relative significance of each of the several factors has varied from sector to sector around the world's sandy coastlines at particular times, and explanations of erosion should be presented in terms of a ranking of these factors for each coastal sector. An example of such an analysis was given by Fisher (1980) from Rhode Island, and is illustrated here in Fig. 93.

SWAMPY COASTLINES

Salt marshes and mangrove swamps have shown rapid advance where the sediment supply has been abundant, in sheltered areas, or on the shores of prograding deltas. Examples of salt marsh progradation in tidal creeks tributary to Chesapeake Bay between 1650 and 1977 were illustrated by Froomer (1980), and similar changes occurred in the Fal estuary between 1878 and 1973 (Ranwell, 1974): in both cases the rear of the salt marsh has been colonized by scrub and woodland vegetation (Fig. 40). In many places the seaward margins of salt marshes and mangrove swamps have retreated in recent decades, so that they terminate in small cliffs fronted by tidal flats. Such salt marsh recession has been noted in southern England and western France, and also in New England, and in these cases the explanation may be a rise of sea level, augmented by land subsidence (Guilcher, 1981). This is certainly true of the salt marshes of Venice Lagoon, which are eroding in response to a rapid rise of sea level relative to the land since the beginning of the present century: changes here can be assessed with reference to photographs taken from the airship *Parsifal* in 1913. Embanking and reclamation of tidal marshlands, which artificially prograde the coastline, have been extensive in north-western Europe, in parts of North America, especially California and New England, and locally elsewhere. In south-east Asia, particularly Indonesia, Thailand and the Philippines, many mangrove areas have been enclosed and converted into brackish-water fishponds.

ARTIFICIAL COASTLINES

Artificialization of coastlines has become extensive during the past century (Walker, 1981): it will be mentioned only briefly here, because the topic is dealt with at length in a sequel to this book, *Artificial Structures on Coastlines*, to be published next year. Sea walls and other structures (Fig. 94) have been built to stop the erosion of cliffs, beaches and delta coastlines. Many sand and shingle spits, previously variable in configuration, have been stabilized by the construction of sea walls or boulder fringes: examples of such armoured spits include Ediz Hook and Sandy Hook in the United States,

Fig. 94 An artificially stabilized coastline at Dymchurch, south-eastern England.
Photo: Eric Bird (November 1961)

and Dawlish Warren in south-west England. Breakwaters have been built to enclose harbours and marinas, which are now widespread in North America, Europe and the Mediterranean, South Africa, Australia, New Zealand and Japan. Large areas have been reclaimed for ports and industrial and urban development alongside estuaries, lagoons and bays. Land reclamation has also been extensive on the coasts of the Netherlands, Singapore, China and Hong Kong, and Tokyo Bay. The spread of artificial structures around the coastline of Port Phillip Bay is indicated on Fig. 74. In Britain, episodes of reclamation of salt marshes and tidal flats have advanced the coastline of The Wash (Kestner, 1962), and the margins of many estuaries, including Poole Harbour (May, 1969) have been modified by the building of sea walls to enclose and reclaim former tidal lands. On the east coast of the Caspian Sea a dam has been built across the mouth of the Kara-Bogaz-Gol, a bordering embayment within which there were high evaporation losses, producing a compensating flow of sea water in from the Caspian. The halting of this flow may be responsible for the recent recovery of levels in the Caspian Sea shown in Fig. 49.

The artificial emplacement and restoration of resort beaches, such as Waikiki Beach at Honolulu, Hawaii, has become extensive in the past two decades in North America and elsewhere. The prevalence of erosion of sandy coasts has posed many problems for resorts: the alternative problem of 'excessive sand', as at Seaside, Oregon, and Malindi, Kenya, has been a rare one. The principle that development of roads and buildings immediately behind sandy coasts is almost invariably followed by an 'erosion problem' necessitating the construction of progressively more elaborate and expensive—and unsightly—structures on the shore has still not been universally understood by coastal planners and developers. The lesson that sea walls lead to refraction scour and accentuate beach depletion has apparently to be learned individually and locally by each coastal engineer working for seaside districts and municipalities.

An alternative to anti-erosion structures is artificial beach nourishment which has been used to offset beach erosion in such places as Bournemouth in England, Port Phillip Bay and the Gold Coast in Australia, Singapore, and at Sochi and Odessa on the Soviet Black Sea Coast. It seems likely that it will also be used as a means of halting cliff erosion, in preference to the traditional, but more artificial, sea walls, boulder heaps, and tetrapods.

An understanding of coastline changes—where and why they have taken place, and where and why they are likely to occur in the future—is essential for planning and development in coastal environments. Fortunately, many people concerned with coastal management are beginning to appreciate the nature of coastline changes, and to seek environmentally acceptable means of dealing with the various problems that arise on changing coastlines.

Bibliography

Ahlmann, H. W. (1949) *Glaciological Research on the North Atlantic Coasts*, Royal Geographical Society, Research Series, 1.
Ahmad, E. (1972) *Coastal geomorphology of India*, Orient Longman, New Delhi.
Alestalo, J. (1985) Finland, in Bird and Schwartz (1985).
Alexander, C. S. (1966) A method of descriptive shore classification and mapping as applied to the north-east coast of Tanganyika, *Ann. Assoc. Am. Geogr.*, **56**, 128–40.
Alexander, C. S. (1969) Beach ridges in north-eastern Tanzania, *Geogr. Review*, **59**, 104–22.
Alexander, C. S. (1985) Hispaniola, in Bird and Schwartz (1985).
Andrews, J. T. (1973) A geomorphological study of post-glacial uplift with particular reference to Arctic Canada, *Inst. Brit. Geogr.*, *Sp. Pub.* 2.
Araya-Vergara, J. F. (1982) Analisis de la localizacion de los procesos y formas predominantes de la llinea litoral de Chile: observacion preliminas, *Inform. Geogr. Chile*, **29**, 35–55.
Araya-Vergara, J. F. (1985) Chile, in Bird and Schwartz (1985).
Armon, J. W. (1974) Late Quaternary shorelines near Lake Ellesmere, Canterbury, *New Zealand Sci. Review*, **10**, 41–5.
Armon, J. W. and McCann, S. B. (1977) Longshore sediment transport and a sediment budget for the Malpleque barrier system, southern Gulf of St. Lawrence, *Canadian J. Earth Sci.*, **14**, 2429–39.
Åse, L. E. (1970) *Shore displacement in eastern Svealand and Åland during the last 4,000 years*, Medd. Naturgeogr. Inst. Stockholms Universitet, A31.
Ayon, H. and Jara, W. (1985) Ecuador, in Bird and Schwartz (1985).
Baines, G. B. K. and McLean, R. F. (1976) Sequential studies of hurricane deposit evolution at Funafuti Atoll, *Mar. Geol.*, **21**, M1–M8.
Ballard, P. (1983) *An ephemeral tombolo at Gabo Island, Victoria,* Royal Military College, Duntroon, Canberra, Department of Geography, Occ. Paper 34.
Baltzer, F. (1969) Les formations vegetales associées au delta de la Dumbéa, *Cahiers ORSTOM*, ser géol., **1**, 59–84.
Banks, R. S. (1975) Beach erosion along the lower west coast of peninsular Florida, *Trans. Gulf Coast Assoc. Geol. Soc.*, **25**, 391–2.
Battistini, R. and Bergoeing, J. P. (1983) *Réconnaissance Géomorphologique de la facade Pacifique du Costa Rica.* Géomorphologie Littorale, CEGET 49.
Battistini, R. and Le Bourdiec, P. (1985) Madagascar, in Bird and Schwartz (1985).
Bedi, N. and Vaidyanadhan, R. (1982) Effect of neotectonics on the morphology of the Narmada River in Gujarat, Western India, *Z. Geomorphol*, **26**, 87–102.
Beke, C. T. (1834) On the historical evidence of the advance of the land on the sea at the head of the Persian Gulf, *Phil. Mag.*, Ser. **3** (6), 401–8.
Berry, R. W., Brophy, G. P., and Naqash, A. (1970) Suspended sediment in the rivers of Iraq, *J. Sed. Petrol.*, **40**, 131–9.
Bird, E. C. F. (1960) The formation of sand beach ridges, *Australian J. Sci.*, **22**, 349–50.
Bird, E. C. F. (1969) The deltaic shoreline near Cairns, Queensland, *Australian Geogr.*, **11**, 138–47.
Bird, E. C. F. (1974) *Coastal changes in Denmark during the past two centuries*, Skrifter i fysisk geografi, 8, Laboratoriet for fysisk geografi, Aarhus Universitet.

Bird, E. C. F. (1976) *Shoreline Changes During the Past Century*, I.G U. Working Group on the Dynamics of Shoreline Erosion, Melbourne.

Bird, E. C. F. (1977) Izmeneniya ochertanyi beregegovoy linii ore zemnogo shara za posledneyo stoletie, *Academy of Sciences USSR Geographical Series*, **3**, 113–9.

Bird, E. C. F. (1978) *A geomorphological study of the Gippsland Lakes Region*. Ministry for Conservation, Victoria, 186.

Bird, E. C. F. (1979) Coastal processes, in K. J. Gregory and D. E. Walling, *Man and Environmental Processes*, Dawson, Folkestone, pp. 82–101.

Bird, E. C. F. (1980) Historical changes on sandy shorelines in Victoria, *Proc. Roy. Soc. Victoria*, **91**, 18–30.

Bird, E. C. F. (1981a) Recent changes on the world's sandy shorelines in E. C. F. Bird and K. Koike, *Coastal Dynamics and Scientific Sites*, Komazawa University, Tokyo, pp. 5–30.

Bird, E. C. F. (1981b) Beach erosion problems at Wewak, Papua New Guinea, *Singapore J. Trop Geogr.*, **2**, 9-14

Bird, E. C. F. (1981c) World-wide trends in sandy shoreline changes during the past century, *Géographie Physique et Quaternaire*, **35**, 241–4.

Bird, E. C. F. (1983) Factors influencing beach erosion and accretion: a global review, in A. MacLachlan and T. Erasmus (eds.) *Sandy Beaches as Ecosystems*, Dr W. Junk, The Hague, pp. 709–717.

Bird, E. C. F. (1984) *Coasts*, Australian National University Press, Canberra and Blackwell, Oxford.

Bird, E. C. F. (1985) Philippines, in Bird and Schwartz (1985).

Bird, E. C. F. and Barson, M. M. (1975) Shoreline changes at Westernport Bay, *Proc. Roy. Soc. Victoria*, **87**, 15–28.

Bird, E. C. F. and Christiansen, C. (1982) Coastal progradation as a by-product of human activity; an example from Hoed, Denmark, *Geogr. Tidsskr.*, **82**, 1–4.

Bird, E. C. F., Dubois, J. P., and Iltis, J. A. (1984) *The impacts of open-cast mining on the rivers and coasts of New Caledonia*. United Nations University.

Bird, E. C. F. and Guilcher, A. (1982) Observations préliminaires sur les récifs frangeants actuels du Kenya et sur les formes littorales associées, *Rev. Géom. Dyn.*, **31**, 113–35.

Bird, E. C. F. and May, V. J (1976) *Shoreline changes in the British Isles during the past century*. Bournemouth Coll. Tech., Division of Geography.

Bird, E. C. F. and Ongkosongo, O. S. R. (1980) *Environmental changes on the coasts of Indonesia*. United Nations University.

Bird, E. C. F. and Paskoff, R. (1979) Relationships between vertical changes of land and sea level and the advance and retreat of coastlines, in K. Suguio (ed.) *Coastal Evolution in the Quaternary*, University of São Paolo, Brazil, pp. 29–40.

Bird, E. C. F. and Ramos, V. T. (1985) Peru, in Bird and Schwartz (1985).

Bird, E. C. F. and Rosengren, N. (1984) Vulcanicity and coastal geomorphology in the Krakatau Islands, *Z. Geomorph.*, **28**, 258–70.

Bird, E. C. F. and Schwartz, M. L. (eds.) (1985) *The World's Coastline*. Van Nostrand Reinhold, Stroudsburg, Pennsylvania.

Bird, J. B., Richards, A. and Wong, P. P. (1979) Coastal subsystems of Western Barbados, West Indies, *Geogr. Annaler*, **61A**, 221–236.

Blanc, J. J. and Grissac, A. J. (1978) *Recherches de géologie sedimentaire sur les herbiers à Posidonies du littoral de la Provence*. CNEXO Marseille.

Bodéré, J. C. (1979) Le rôle essential des débacles glacio-volcaniques dans l'évolution recente des côtes sableuses en voie de progradation du sud-est d'Islande, in A. Guilcher (ed.) *Les Côtes Atlantiques de l'Europe*, University of Western Brittany, Brest, pp. 55–64.

Bodéré, J. C. (1985) Iceland, in Bird and Schwartz (1985).

Borówca, R. K. (1985) Poland, in Bird and Schwartz (1985).

Bourman, R. P. (1974) Historical geomorphic change: Fleurieu Peninsula, South Australia, *Proc. I.G.U. Regional Conf.*, Palmerston North, New Zealand, 289–97.

Bourman, R. P. (1976) Environmental geomorphology, *Proc. Roy. Geogr. Soc. Australasia, S. Australian Branch*, **76**, 1–23.

Bremner, J. M. (1985) South-west African/Namibia, in Bird and Schwartz (1985).

Briquet, A. (1930) *Le littoral du Nord de la France*, The University of Paris.

Broggi, J. A. (1946) Geomorfologica del delta Rio Tumbes, *Soc. Geog. de Lima*, Boletin 24.

Brown, M. J. F. (1974) A development consequence: disposal of mining waste on Bougainville, Papua New Guinea, *Geoforum*, **18**, 19–27.

Brunsden, D. and Jones, D. K. C. (1980) Relative time scales and formative events in coastal landslide systems, *Z. Geomorph.*, Supp. **34**, 1–19.

Bruun, P. (1962) Sea level rise as a cause of shore erosion, *J. Waterways & Habors Div., Proc. Amer. Soc. Civil Engnrs*, **88**, 117–30.

Bryan, R. B. and Price, A. G. (1980) Recession of the Scarborough Bluffs. Ontario, Canada, *Z. Geomorph.*, Supp. **34**, 48–62.

Bryant, E. A. (1979) Wave climate effects upon changing barrier island morphology, Kouchibouguac Bay, New Brunswick, *Maritime Sediments*, **15**, 49–62.

Bryant, E (1983a) Coastal erosion and accretion, Stanwell Park Beach, N.S.W., *Australian Geogr.*, **15**, 382–390.

Bryant, E. (1983b) Regional sea level, southern oscillation and beach change, New South Wales, Australia, *Nature*, **305**, 213–216.

Byrne, J. V. (1964) An erosional classification for the northern Oregon coast, *Annals Assoc. Amer. Geogr.*, **54**, 329–35.

Caldwell, J. M. (1966) Coastal processes and beach erosion, *J. Boston Soc. Civil Engnrs*, **53**, 142–57.

Calles, B., Lindé, K., and Norrman, J. O. (1982) The geomorphology of Surtsey Island in 1980. *Surtsey Research Progress Report*, **9**, 117–132.

Cambers, G. (1976) Temporal scales in coastal erosion systems, *Trans. Inst. Brit. Geogr.*, NS **1**, 246–256.

Cameron, H. L. (1965) The shifting sands of Sable Island, *Geogr. Rev.*, **44**, 363–76.

Campbell, J. F. (1972) *Erosion and accretion of selected Hawaiian beaches, 1962–1972*. University of Hawaii.

Campbell, J. F. and Hwang, D. J. (1982) Beach erosion at Waimea Bay, Oahu, Haiwaii, *Pacific Science*, **36**, 35–43.

Caputo, C., D'Allesandro, L., La Monica, G. B., Landini, B., Lupia Palmieri, E., and Pugliese, F. (1983) Erosion problems on the coast of Latium, Italy, in E. C. F. Bird and P. Fabbri (eds.) *Coastal Problems in the Mediterranean Sea*, University of Bologna, Italy, pp .59–68.

Carr, A. P. (1962) Cartographic record and historical accuracy, *Geography*, **47**, 135–44.

Carr, A. P. (1969) The growth of Orford spit: the cartographic and historical evidence from the sixteenth century, *Geogr. J.* **135**, 28–39.

Carr, A. P. and Gleason, R. (1972) Chesil Beach and the cartographic evidence of Sir John Coode, *Proc. Dorset Nat. Hist. Archaeol. Soc.*, **93**, 125–131.

Carter, C. H. (1978) *Lake Erie shore erosion*. Ohio Geol. Surv. Report 99.

Carter, R. W. G. (1979) Recent progradation of the Magilligan Foreland, County Londonderry, in A. Guilcher (ed.) *Les Côtes Atlantiques de l'Europe*, University of Western Brittany, Brest, pp. 17–28.

Carter, R. W. G. (1980) Human activities and coastal processes: the example of recreation in Northern Ireland, *Z. Geomorph.*, Supp. **34**, 155–64.

180

Carter, R. W. G., Lowry, P., and Shaw, J. (1983) An eighty year history of erosion in a small Irish bay, *Shore and Beach,* **34**, 34–39.

Cencini, C., Cuccoli, L., Fabbri, P., Montanari, F., Sembolini, F., Torresani, S. and Varani, L. (1979) *Le spiagge di Romana: Uno spazzio da proteggere.* Con. Naz. delle Ricerche, Bologna.

Chamberlain, T. (1968) The littoral sand budget, Hawaiian Islands, *Pacific Sci.,* **22**, 161–83.

Chambers, M. J. G. and Sobur, A. J (1975) *The rates and processes of recent coastal accretion in the province of South Sumatra.* I.P.B., Bogor.

Chapman, D. M., Geary, M., Roy, P. S. and Thom, B. G. (1982) *Coastal evolution and coastal erosion in New South Wales.* Coastal Council of N.S.W., Sydney.

Chappell, J. (1974) Geology of coral terraces, Huon Peninsula: a study of Quaternary tectonic movements and sea-level changes, *Bull. Geol. Soc. America,* **85**, 553–70.

Chen Jiyu, Liu Cangzi, and Yu Zhiying (1985) China, in Bird and Schwartz (1985).

Christiansen, C., Christoffersen, H., Dalsgaard, J., and Nornberg, P. (1981) Coastal and nearshore changes correlated with die-back in eel grass (Zostera marina), *Sedim. Geol.,* **28**, 163–73.

Christiansen, C. and Møller, J. T. (1980) Beach erosion at Klim, Denmark: a ten year record, *Coastal Engineering,* **3**, 283–96.

Clayton, K. M. (1980) Coastal protection along the East Anglian coast, *Z. Geomorph.,* Supp. **34**, 163–172.

Coakley, J. P. (1976) The formation and evolution of Point Pelée, western Lake Erie, *Canadian J. Earth Sci.,* **13**, 136–44.

Craig, A. K. (1985) Bahamas, in Bird and Schwartz (1985).

Craig, A. K. and Psuty, N. P. (1968) The Paracas Papers, *Studies in Marine Desert Ecology,* **1** (2), 44–77.

Cruz, O., Coutinho, P. N., Duarte, G. M., Gomes, A. and Muehe, D. (1985) Brazil, in Bird and Schwartz (1985).

Daetwyler, C. C. (1965) *Marine geology of Tomales Bay, Central California.* Scripps Inst. of Oceanography.

Davies, J. L. (1957) The importance of cut and fill in the development of beach sand ridges, *Australian J. Sci.,* **20**, 105–11.

Davies, J. L. (1985) Tasmania, in Bird and Schwartz (1985).

Deane, C. A. W. (1985) Lesser Antilles, in Bird and Schwartz (1985).

De Boer, G. (1969) The historical variations of Spurn Head: the evidence of early maps, *Geogr. J.,* **135**, 17–27.

De Boer, G. (1978) Holderness and its coastal features, in D. Symes (ed.) *North Humberside, Introductory Essays,* University of Hull, pp. 69–76.

Dei, L. A. (1972) The central coastal plains of Ghana: a morphological and sediment-ological study, *Z. Geomorph.,* **16**, 415–31.

Del Canto, S. and Paskoff, R. (1983) Características y evolucíon geomorfológica actual de algunas playas de Chile central, entre Valparaíso y San Antonio, *Revista de Geografía Norte Grande,* **10**, 31–45.

Depuydt, F. (1972) De Belgische strand, *Verh. Kon. Akad. Wetensch.,* 122.

Dionne, J. C. (1972) Caracteristiques des schorres des regions froides, en particulier de l'éstuaire du Saint-Laurent, *Z. Geomorph.,* **13**, 131–162.

Dionne, J. C. (1979) *L'érosion des rives du Saint-Laurent: une menace sérieuse a l'environnement.* Environnement Canada.

Diresse, P. and Kouyoumontzakis, G. (1985) Gabon, Congo, Cabinda and Zaire, in Bird and Schwartz (1985).

Dolan, R. and Bosserman, K. (1972) Shoreline erosion and the lost colony, *Ann. American Assoc. Geogr.* **62**, 424–426.

Dolan, R., Hayden, B. P., May, P. and May, S. (1980). The reliability of shoreline change measurements from aerial photographs, *Shore and Beach,* **48**, 22–29.

Dolan, R. Hayden, B., and Heywood, J. (1978). Analysis of coastal erosion and storm surge hazards, *Coastal Engineering*, **2**, 41–53.

Dolan, R., Hayden, B., Rea, C. and Heywood, J. (1980) Erosion at Cape Lookout: will the lighthouse fall?, *Shore and Beach*, **48**, 13–20.

Dresch, J. (1961) Observations sur le desert côtier du Perou, *Ann. de Gogr.* **70**, 179–84.

Dubois, J. M. (1973) *Etude des rives du St-Laurent*. Ministère des Travaux Public, Canada.

Dubois, J. M. M. (1980) Géomorphologie du littoral de la côte du Saint-Laurent: analyse sommaire, in S. B. McCann (ed.) *The Coastline of Canada*, Geological Survey of Canada, Paper 80–10, pp. 215–38.

Dunbar, G. S. (1956) *Geographical history of the Carolina banks*, Coastal Studies Institute, Louisiana State University, Tech. Report 8.

Edelman, T. (1977) Systematic measurements along the Dutch coast, *Proc. 10th Conf. Coastal Engineering*, pp. 489–501.

Eisma, D. (1962) Beach ridges near Selçuk, Turkey, *Tijdschr. K. Ned. Aardrijksk Genocot.*, **79**, pp. 234–45.

Eisma, D. (1985) Vietnam, in Bird and Schwartz (1985).

Eisma, D. and Park, D. W. (1985) North and South Korea, in Bird and Schwartz (1985).

El Ashry, M. T. (ed.) (1977) *Air Photography and Coastal Problems*. Dowden, Hutchinson and Ross, Stroudsburg, USA.

Eliot, I. G. Clarke, D. J., and Rhodes, A. (1982) Beach-width variation at Scarborough, Western Australia, *J. Roy. Soc. W. Australia*, **65**, 153–8.

Ellenberg, L. (1985) Venezuela, in Bird and Schwartz (1985).

Emery, K. O. (1960) *The Sea off Southern California*, Wiley, New York.

Emery, K. O. and Neev, D. (1960) Mediterranean beaches of Israel, *Min. Devt. Jerusalem, Geol. Surv. Bulletin*, **26**, 1–24.

Erol, O. (1983) Historical changes on the coastline of Turkey, in E. C. F. Bird and P. Fabbri (eds.), *Coastal Problems of the Mediterranean Sea*, University of Bologna, Italy, pp. 95–107.

Erol, O. (1985) Turkey, in Bird and Schwartz (1985).

Evans, G. (1971) The recent sedimentation of Turkey, in A. S. Campbell (ed.), *Geology and History of Turkey*, Petroleum Exploration Society of Libya, Tripoli, pp. 385–406.

Everard, C. E. (1962) Mining and shoreline evolution near St. Austell, Cornwall, *Trans. Roy. Geol. Soc. Cornwall*, **19**, 199–219.

Facon, R. (1965) La pointe de la Coubre, etude morphologique, *Norois*, **12**, 165–80.

Fairbridge, R. W. (1966) Mean sea level changes, *Encyclopaedia of Oceanography*, Reinhold, New York, pp. 479–85.

Fels, E. (1944) Landgewinnung in Griechenland, *Petermanns, Geogr. Mitt.*, 242.

Fisher, J. J. (1980) Shoreline erosion, Rhode Island and North Carolina coasts, in M. L. Schwartz and J. J. Fisher (eds.), *Proceedings of the Per Bruun Symposium*, University of Rhode Island, Newport, pp. 32–54.

Fisher, J. J. (1985) Atlantic U.S.A—North, in Bird and Schwartz (1985).

Fisher, J. J. and Regan, D. R. (1978) Determination of precision of photogrammetric measurements for changes of shorelines in W. F. Tanner (ed.) *Standards for Measuring Shoreline Changes*, Florida State University, Tallahassee, pp. 32–54.

Fisher, J. J. and Simpson, E. J. (1979) Washover and tidal sedimentation rates as environmental factors in development of a transgressive barrier shoreline, in S. P. Leatherman (ed.), *Barrier Islands from the Gulf of St. Lawrence to the Gulf of Mexico*, Academic Press, New York, pp. 127–48.

Froomer, N. (1980) Morphologic changes in some Chesapeake Bay tidal marshes, *Z. Geomorph.*, Supp. **34**, 242–54.

Galloway, R. W. (1985). Northern Territory, in Bird and Schwartz (1985).

Gastescu, P. and Breier, A. (1980) Present changes in the Danube delta morphohydrography, *Rev. Roum. Geogr.* **24**, 41–46.

Gibb, J. G. (1978) Rates of coastal erosion and accretion in New Zealand, *New Zealand J. Marine Fresh. Research*, **12**, 429–450.

Gierloff-Emden, H. G. (1969) *Die Küste von El Salvador*. Acta Humboldtiana Series Geog. Ethnog., 2.

Gierloff-Emden, H. G. (1985a) Baltic West Germany, in Bird and Schwartz (1985).

Gierloff-Emden, H. G. (1985b) North Sea, West Germany, in Bird and Schwartz (1985).

Gilbert, G. K. (1917) *Hydraulic mining debris in the Sierra Nevada*. U.S. Geol. Surv. Prof. Paper 105.

Goldsmith, V. (1983) Dynamic geomorphology of the Israeli coast: a brief review, in E. C. F. Bird and P. Fabbri (eds.) *Coastal Problems in the Mediterranean Sea*, University of Bologna, Italy, pp. 109–120.

Goldsmith, V. and Golik, A. (1980) Sediment transport model of the southeastern Mediterranean coast, *Mar. Geol.* **37**, 147–75.

Goldsmith, V., Sutton, C. H. Frish, A., Heiligman, M. and Haywood, A. (1978) The analysis of historical shoreline changes, *Coastal Zone '78*, **4**, 2819–36.

Granö, O. (1981) An emerging esker in southern Finland, *Geogr. Annaler*, **63A**, 293–301.

Griggs, G. B., and Johnson, R. E. (1979) Coastline erosion, Santa Cruz County. *California Geology*, **32**, 67–76.

Gudelis, Y. (1985) Baltic Coast, USSR, in Bird and Schwartz (1985).

Guilcher, A. (1958) *Coastal and Submarine Morphology*. Methuen, London.

Guilcher, A. (1965) Drumlin and spit structures in the Kenmare River, south-west Ireland, *Irish Geogr.*, **5**, 7–19.

Guilcher, A. (1981) Shoreline changes in salt marshes and mangrove swamps (mangals) within the past century, in E. C. F. Bird and K. Koike (eds.) *Coastal Dynamics and Scientific Sites*, Komazawa University, Tokyo, pp. 31–53.

Guilcher, A. (1985) France, in Bird and Schwartz (1985).

Guilcher, A. Medeiros, C. A. Matos, J. E., and Oliveira, J. T. (1974) Les restingas (flèches littorales) d'Angola, *Finistere*, **9**, 171–211.

Guilcher, A. and Nicholas, J. P. (1954) Observations sur la langue de Barbarie et les bras du Sénégal aux environs de St Louis, *Bull. Inf. Com. Cent. Océanogr. Etude Côtes*, **6**, 227–42.

Gulliver, F. P. (1899) Shoreline topography, *Proc. Amer. Acad. Arts, Sci.*, **34**, 151–258.

Hallegouët, B. and Moign, A. (1976) Historique d'une évolution de littoral dunaire: la baie de Goulven (Finistère), *Penn ar Bed.*, **10**, 263–276.

Hallegouët, B. and Moign, A. (1979) Progradation et érosion d'un secteur littoral sableuse en Bretagne Nord: mésures et bilan, in A. Guilcher (ed.) *Les Côtes Atlantiques de l'Europe*, University of Western Brittany, Brest, pp. 45–54.

Harvey, N. (1983) *The Murray mouth*. Geography Teachers' Association of South Australia.

Hayes, M. O. (1985) Atlantic U.S.A.—South, in Bird and Schwartz (1985).

Healy, T. R. (1977) Progradation at the entrance to Tauranga Harbour, Bay of Plenty, *New Zealand Geogr.*, **33**, 90–91.

Heerden, I. L. V. and Roberts, H. H. (1980) The Atchafalaya delta: rapid progradation along a traditionally retreating coast, *Z. Geomorph.*, Supp. **34**, 184–201.

Helle, R. (1965) Strandwallbildungen im gebiet am unterlauf des flusses Siikajoki, *Fennia*, **95**, 5–35.

Hehanussa, P. E. (1979) Excursion guide to the Cimanuk delta complex, West Java, in E. C. F. Bird and A. Soegiarto (eds.) *Proceedings of the Jakarta Workshop on Coastal Resources Management*, United Nations University, Tokyo, pp. 92–104.

Herd, D. G., Yourd, T. L., Hansjurgen, C., Person, W. J., and Mendoza, C. (1981) The great Tumaco, Columbia, earthquake of 12 December 1979, *Science*, **211**, 441–5.

Hernandez Pacheca, F. (1966) Caraceristicas generales del litoral cantábrico y el proceso de relleno de sus rias, *Rev. Inf. INI*, pp. 1–46.

Heusser, C. (1960) Late Pleistocene environments of Laguna de San Rafael, Chile, *Geogr. Rev.*, **50**, 555–7.

Heydorn, A. E. F., and Tinley, K. L. (1970) Estuaries of the Cape, Part I—Synopsis of the Cape Coast, National Research Institute Oceanology, Stellenbosch.

Hicks, S. D. and Shofnos, W. (1965) The determination of land emergence from sea level observations in southeast Alaska, *J. Geophy. Res.*, **70**, 3315–20.

Hinschberger, F. (1985) Ivory Coast, in Bird and Schwartz (1985).

Hodgkin, E. (1976) The history of two coastal lagoons at Augusta, Western Australia, *J. Roy. Soc. W. Australia*, **59**, 39–45.

Höllerwoger, F. (1966) The progress of the river deltas in Java, *Sci. Problems Humid Tropical Zone Deltas*, pp. 347–55.

Hsu, T. L. (1985) Taiwan, in Bird and Schwartz (1985).

Hunter, J. F. (1914) Erosion and sedimentation in Chesapeake Bay, *U.S. Geol. Surv. Prof. Paper*, **40**, 7–15.

Hutchinson, J. N. (1971) Field and laboratory studies of a fall in Upper Chalk cliffs at Joss Bay, Isle of Thanet, *Proc. Roscoe Memorial Symp.* pp. 692–706.

Hutchinson, J. N. (1976) Coastal landslides in cliffs of Pleistocene deposits between Cromer and Overstrand, Norfolk, England, *Laurits Bjerrum Memorial Volume*, University of Oslo, pp. 155–182.

Hwang, D. (1981) *Beach changes on Oahu as revealed by aerial photographs*, Hawaii Institute of Geophysics, HIG-81/3.

Iriondo, M. and Scotta, E. (1979) The evolution of the Parana River delta, *Proc. Int. Symp. Coastal Evol. Quaternary*, pp. 405–18.

Jackson, J. M. (1985) Uruguay, in Bird and Schwartz (1985).

Jennings, J. N. (1963) Some geomorphological problems of the Nullarbor Plain, *Trans Roy. Soc. S. Australia*, **87**, 41–62.

Jennings, J. N. (1975) Desert dunes and estuarine fill in the Fitzroy estuary (N. W. Australia), *Catena*, **2**, 215–62.

Johnson, D. W. (1919) *Shore Processes and Shoreline Development*, Wiley, New York.

Johnson, D. W. (1925) *The New England-Acadian shoreline*, Wiley, New York.

Johnston, W. A. (1921) Sedimentation of Fraser River delta, *Geol. Surv. Canada, Memo.*, p. 125.

Jones, M. (1977) *Finland, daughter of the sea*. Dawson, Folkestone.

Jones, M. (1982) Landhevning, bosettingsmonster, og jordeiendomsforhold i historic lys: et nordisk perspektiv, *Landhöjning och Kustbygdsförändring, Symposiepublikation*, **1**, 235–62.

Kaplin, P. (1981) Relief, age and types of oceanic islands, *New Zealand Geogr.*, **37**, 3–12.

Kaplin, P. (1985) *Pacific coast, USSR*, in Bird and Schwartz (1985).

Kassar, M. (1972). The impact of river control schemes on the shoreline of the Nile delta, in M. T. Farrar and J. P. Milton (eds.) *The Careless Technology*, Natural History Press , New York, pp. 179–188.

Kaye, C. A. (1973) *Map showing changes on shoreline of Martha's Vineyard, Massachusetts, during the past 200 years*. U.S.G.S. Misc. Field Studies, Map MF–534.

184

Kelletat, D. (1972) Beiträge sur regionalen Küstenmorphologie des Mittelmeerraumes Gargana/Italien und Peloponnes/Griechenland, *Z. Geomorphol*, N. F. Supplementband 19.
Kestner, F. J. T. (1962) The old coastline of The Wash, *Geogr. J.*, **128**, 457–78.
Khafagy, A. and Manohar, M. (1979) Coastal protection of the Nile delta, *Nature and Resources*, **15**, 7–13.
King, C. A. M. (1956) The coast of south-east Iceland near Ingolfshöfdi, *Geogr. J.*, **122**, 241–6.
King, C. A. M. (1969) Some arctic coastal features around Foxe Basin and in East Baffin Island, *Geogr. Annaler*, **A–51**, 207–18.
Kirk, R. M. (1975) Coastal changes at Kaikoura, 1942–74, *New Zealand J. Geol. Geophys*, **18**, 787–802.
Klemsdal, T. (1985) Norway, in Bird and Schwartz (1985).
Koike, K. (1977) The recent change of sandy shorelines in Japan, *Komazawa Geogr.*, **13**, 1–16.
Koike, K. (1985) Japan, in Bird and Schwartz (1985).
Komar, P. D. (1985) Oregon, in Bird and Schwartz (1985).
Koopmans, B. N. (1972) Sedimentation in the Kelantan delta, Malaysia, *Sediment, Geol.*, **7**, 65–84.
Kraft, J. C. 1985 Atlantic U.S.A—Central, in Bird and Schwartz (1985).
Kraft, J. C., Aschenbrenner, S. E., and Rapp, G. (1977) Palaeo-geographic reconstruction of coastal Aegean archaeological sites, *Science*, **195**, 941–7.
Kramer, J. (1978) Coast protection works on the German North Sea coast, *Die Küste*, **32**, 124–139.
Kukla, G. and Gavin, J. (1983) Antarctic ice volumes, *Science*, **214**, 497.

La Fond, E. C. (1966) Bay of Bengal, in R. W. Fairbridge (ed.) *Encylopaedia of Oceanography*, Reinhold, New York, pp. 110–18.
Larsen, C. (1973) *Variation in bluff recession in relation to lake level fluctuations*. Illinois Inst. Env. Quality, Chicago.
Le Bourdiec, P. (1958) Aspects de la morphogenèse plio-quaternaire en basse Côte d'Ivoire, *Revue Géomorph. Dyn.*, **9**, 33–42.
Lees, G. M. and Falcon, N. L. (1952) The geographical history of the Mesopotamian Plains, *Geogr. J.*, **18**, 24–39.
Leontiev, O. K. (1985) Caspian U.S.S.R in Bird and Schwartz (1985).
Leontiev, O. K., Maev, E. G., and Rychagov, C. (1977) *Geomorphology of the Caspian Sea*, Moscow State University.
Lewellen, R. (1970) *Permafrost erosion along the Beaufort Sea coast*. University of Denver.
Lindh, G. (1976) Aspects of the beach erosion problem in south Sweden, *Svensk. Geogr. Arsbok.*, **52**, 5–19.
Löffler, E. (1977) *The Geomorphology of Papua New Guinea*. Australian National University Press, Canberra.
Luck, G. (1978) Islands in front of the German North Sea coast, *Die Küste*, **32**, 94–109.
Lustig, T. L. (1977) Aeolian-induced cyclic meandering of an ephemeral deltaic river mouth, *3rd Australian Conf. Coastal Ocean Engineers*, pp. 234–9.
Luternauer, J. L. (1980) Genesis of morphologic features on the western delta front of the Fraser River, British Columbia, in S. B. McCann (ed.) *The Coastline of Canada*, 381–96. Geological Survey of Canada, Ottawa, Paper 80–10, pp. 381–96.

MacCarthy, G. R. (1953) Recent changes in the shoreline near Point Barrow, Alaska, *Arctic*, **6**, 44–51.
Macdonald, H. V., Robinson, D. A., and Clark, J. L. (1973) Beach erosion investigations in Queensland, *Inst. Engineers Australia National Conference*, **73**(1), 21–8.

Mackay, J. R. (1963a) *The Mackenzie delta.* Geogr. Branch, Mines & Tech. Surveys, Ottawa, Memoir.

Mackay, J. R. (1963b) Notes on shoreline recession along the coast of the Yukon Territory, *Arctic*, **16**, 195–7.

Mahrour, M. and Dagorne, A. (1985) Algeria, in Bird and Schwartz (1985).

Mamykina, V. A. (1978) Recent processes in the coastal zone of the Azov Sea, *Proc. Geogr. Soc. U.S.S.R.*, **110**, 351–9.

Manley, G. (1952) *Climate and the British Scene.* Collins, London.

Marker, M. E. (1967) The Dee estuary: its progressive silting and salt marsh development, *Trans. Inst. Brit. Geog.*, **41**, 65–71.

Marques, M. A. and Julia, R. (1983) Coastal problems in Alt Emporda, Cataloñia, in E. C. F. Bird and P. Fabbri (eds.) *Coastal Problems in the Mediterranean Sea* University of Bologna, Italy, pp. 83–94.

Marshall, P. (1933) Effects of earthquake on coastlines near Napier, *New Zealand J. Sci. Tech.*, **15**, 79–92.

Martin, L., Bittencourt, A. C. S. P., Villas Boas, G. B., and Flexor, J. M. (1980) *Mapa geologico do Quaternaireio Costeiro do Estado da Bahia.*

Martini, I. P. (1981) Morphology and sediments of the emergent Ontario coast of James Bay, Canada, *Geogr. Annaler*, **63A**, 81–94.

Mathlouti, S. and Paskoff, R. (1981) Modifications de la ligne de rivage dans la Baie de Bizerte depuis un siècle, *Revue Tunisienne de Géogr.*, **24**, 91–103.

May, S. K., Kimball, W. H., Grandy, N., and Dolan, R. (1982) The Coastal Erosion Information System, *Shore and Beach*, **50**, 19–26.

May, V. J. (1964) Reclamation and shoreline change in Poole Harbour, Dorset, *Proc. Dorset Nat. Hist. Archaeol. Soc.*, **90**, 141–54

May, V. J. (1971) The retreat of chalk cliffs, *Geogr. J.*, **137**, 203–6.

May, V. J., (1979) Changes on the coastline of southwest England – a review, in A. Guilcher (ed.) *Les Cotes Atlantiques de l'Europe*, University of Western Brittany, Brest, pp. 65–76.

Mazzanti, R. and Pasquinucci, M. (1983) The evolution of the Luni-Pisa coastline, in E. C. F. Bird and P. Fabbri (eds.) *Coastal Problems in the Mediterranean Sea*, University of Bologna, Italy, pp. 47–58.

McCann, S. B. (1973) Beach processes in an Arctic environment, in D. R. Coates (ed.) *Coastal Geomorphology*, State University of New York, Binghamton, pp. 141–55.

McCann, S. B. and Hale, P. B. (1980) Sediment dispersal patterns and shore morphology along the Georgia Strait coastline of Vancouver Island, *Proc. Canadian Coastal Conference*, Coastal Society, Burlington, Ontario, pp. 151–163.

McCormick, C. L. (1973) Probable causes of shoreline recession and advance on the south shore of eastern Long Island, in D. R. Coates (ed.) *Coastal Geomorphology*, State University of New York, Binghamton, pp. 61–75.

McIntire, W. G. and Walker, H. J., (1964) Tropical cyclones and coastal geomorphology in Mauritius, *Ann. Assoc. American Geogr.* **54**, 582–96.

McLean, R. F. (1978) Recent coastal progradation in New Zealand, in J. L. Dvies and M. A. J. Williams (eds.) *Landscape Evolution in Australasia*, Australian National University, Canberra, pp. 168–196.

McLean, R. F. (1979) The coast of Lakeba, a geomorphological reconnaissance, in H. C. Brookfield (ed.) *Fiji Island Reports*, UNESCO/UNFPA-MAB Programme, Canberra, **5**, 67–81.

Meijerink, A. M. J. (1982) Dynamic geomorphology of the Mahanadi delta, *I.T.C. Journal, Special Verstappen Issue*, pp. 243–250.

Meinesz, A. and Lefevre, J. R. (1978) Destruction de l'étage infralittoral des Alpes Maritimes (France) et de Monaco par les restructurations du rivage. *Bulletin D'écologie*, **9**, 259–76.

Meinesz, A., Astier, J. M. and Lefevre, J. R. (1981) Impact de l'aménagement du domaine maritime sur l'étage infralittoral du Var, France (Mediterranée Occidentale). *Annales de l'Institut Océanographique,* **57**, 65–77.

Meinesz, A., Astier, J. M. Bodoy, A., Cristiani, G., and Lefevre, J. R. (1983a) Impact de l'aménagement du domaine sur l'étage infralittoral des Bouches-du-Rhône (Mediterranée Occidentale). *Vie et Milieu,* **32**, 101–10.

Meinesz, A., Lefevre, J. R. Boudouresque, D. F., Beurrier, J. P., Miniconi, R., and O'Neilly, P. (1983b) Les zones marines protegées des côtes francaises de la Mediterranée. *Bulletin d'Écologie,* **14** (1).

Miller, D. J. (1960) Giant waves in Lituya Bay, Alaska, *U.S. Geol. Soc. Prof. Pap.* **354–C**, 51–86.

Miossec, J. and Paskoff, R. (1979) Evolution des plages et aménagements touristiques à Jerba (Tunisie), *Mediterranée,* **1**, 99–106.

Moberly, R. and Chamberlain, T. (1964) *Hawaiian beach system.* Hawaii Institute of Geophysics, Report 64–2.

Moign, A. and Guilcher, A. (1967) Une flèche littorale en milieu périglaciaire arctique: la fléche de Sars (Spitsberg), *Norois,* **56**, 549–68.

Moldovanu, A. and Selariu, O. (1971) Contributions to the study of geomorphological processes in the shore zone of the Black Sea between Constanta and Agigea, *Studii si Cercetari Geograf. Aplic. Dobregei,* pp. 77–83 (in Rumanian).

Molnia, B. F. (1977) Rapid shoreline erosion at Icy Bay, Alaska, *Proc. 11th Offshore Technology Conference,* **4**, 115–26.

Molnia, B. F. (1979) Sedimentation in coastal embayments, north eastern Gulf of Alaska, *Proc. 9th Offshore Technology Conference,* pp. 665–70.

Møller, J. T. (1963) Vadehavet Mellem og Ribe Å, *Folia Geogr. Danica,* **8** (4).

Morelock, J. (1978) *Shoreline of Puerto Rico.* Department of Natural Resources, Puerto Rico.

Morelock, J. and Trumbull, J. V. A. (1985) Puerto Rico, in Bird and Schwartz (1985).

Morgan, J. P. (1963) Changes on the Louisiana shoreline, *Louisiana State University Coastal Studies Institute,* Contribution 63–5, pp. 66–78.

Morton, R. A. (1977) Historical shoreline changes and their causes, Texas Gulf coast, *Texas Gulf Coast Assoc. Geol. Soc.,* **27**, 352–64.

Mroczek, P. (1980) Zu einer Karte der Veränderungen der Uferlinie der deutschen Nordseeküste in den letzen 100 Jahren, *Berliner Geogr. Stud.,* **7**, 39–57.

Muehe, D. (1979) Sedimentology and topography of a high energy coastal environment between Rio de Janeiro and Cabo Frio, *Ann. Acad. Bras. Ciencias,* **51**, 473–81.

Nagaraja, V. N. (1966) Hydrometeorological and tidal problems of the deltaic areas in India, *Sci. Problems Humid Tropical Zone Deltas:* pp. 115–19.

Nageswara Rao, K. and Vaidyanadhan, R. (1979) Evolution of the coastal landforms on the Krishna delta front, India, *Trans. Inst. Indian Geogr.,* **1**, 25–32.

Nielsen, E. (1973) Coastal erosion in the Nile delta, *Nature and Resources,* 1973/N1, pp. 14–18.

Nielsen, N. (1969) Morphological studies on the eastern coast of Disko, West Greenland, *Geografisk Tidsskrift,* **68**, 1–35.

Nielsen, N. (1985) Greenland, in Bird and Schwartz (1985).

Nir, Y. (1985) Israel, in Bird and Schwartz (1985).

Nordstrom, K. F. and Allen, J. R. (1980) Geomorphically compatible solutions to beach erosion, *Z. Geomorph.,* Supp. **34**, 142–54.

Norris, R. M. (1952) Recent history of a sand spit at Saint Nicolas Island, California, *J. Sed. Petrol.,* **22**, 224–8.

Norrman, J. O. (1980) Coastal erosion and slope development in Surtsey Island, Iceland, *Z. Geomorphol.,* Supp. **34**, 20–38.

Nossin, J. J. (1964) Geomorphology of the surroundings of Kuantan Eastern Malaya, *Geol. en Mijnbouw*, **43**, 157–82.

Nossin, J. J. (1965) The geomorphic history of the northern Padang delta, *J. Trop. Geogr.*, **20**, 54–64.

Nummedal, D. and Stephen, M. F. (1976) *Coastal dynamics and sediment transportations, northeast Gulf of Alaska*. Univ. South Carolina, Tech. Rep. 9-CRD.

Oertel, G. F. and Chamberlain, C. F. (1975) Differential rates of shoreline advance and retreat of coastal barriers, Georgia, *Trans. Gulf Coast Assoc. Geol. Socs.*, **25**, 383–90.

Ogden, J. G. (1972) Erosion of island beaches linked to rise of sea level, *Martha's Vineyard Gazette*, 22 September 1972.

Olson, J. S. (1958) Lake Michigan dune development, *J. Geol.*, **66**, 254–63, 345–51, and 473–83.

Orford, J. D. and Carter, R. W. G. (1982) Recent geomorphological changes on the barrier coasts of south-east County Wexford, *Irish Geogr.*, **15**, 70–84.

Orlova, G. and Zenkovich, V. P. (1974) Erosion on the shores of the Nile delta, *Geoforum*, **18/74**, 68–72.

Orme, A. R. (1973) Barrier and lagoon systems along the Zululand coast, South Africa, in D. R. Coates (ed.) *Coastal Geomorphology*, State University of New York, Binghamton, pp. 182–212.

Orme, A. R. (1985) California, in Bird and Schwartz (1985).

Ottmann, F. (1965) *Introduction á la Géologie Marine et Littorale*, Masson, Paris.

Owens, E. H. (1974) A framework for the definition of coastal environments in the southern Gulf of St Lawrence, *Geol. Surv. Canada Paper*, **74/30**, 47–76.

Owens, E. H. and Bowen, A. J. (1977) Coastal environments of the Maritime Provinces, *Maritime Sediments*, **13**, 1–32.

Owens, E. H. and Harper, J. R. (1985) British Columbia, in Bird and Schwartz (1985).

Ozasa, H. (1977) Recent shoreline changes in Japan—an investigation using aerial photographs, *Coastal Engineering in Japan*, **20**, 69–81.

Parker, R. (1978) *Men of Dunwich*, Collins, London.

Paskoff, R. (1978) Aspects géomorphologiques de l'île de Pacques, *Bull. Assoc. Géogr. Franc.*, **452**, 142–57.

Paskoff, R. (1978), L'évolution de l'embouchure de la Medjerda, *Photo-Interprétation*, **5**, 1–23.

Paskoff, R. (1981a) *L'érosion des côtes*, Presses Universitaries de France, Paris.

Paskoff, R. (1981b) Evolution recente de la fléche de Foum el Qued, delta de la Medjerda, *Mediterranée*, **4**, 39–42.

Paskoff, R. (1981c) L'érosion des plages en Tunisie, *Revenue Tunisienne de Géographie*, **8**, 81–96.

Paskoff, R. (1983) L'érosion des plages, *La Recherche*, **140**, 20–8.

Paskoff, R. and Sanlaville, P. (1978) Observations géomorphologiques sur les côtes de l'archipel maltais, *Z. Geomorph.*, **22**, 310–28.

Petersen, M. (1978) Islands along the eastern North Sea coast, *Die Küste*, **32**, 124–39.

Pichler, H., Gunther, D., and Kussmaul, S. (1972) Inselbildung and magmengenese in Santorin, *Naturwissenschaften*, **19**, 188–97.

Pierce, J. W. (1969) Sediment budget along a barrier island chain, *Sediment, Geol.*, **3**, 5–16.

Piper, D. J. W., and Panagos, A. G. (1981) Growth patterns of the Acheloos and Evinos deltas, *Sedim Geol.*, **28**, 111–32.

Pirazzoli, P. A., Thommeret, J., Thommeret, Y., Laborel, J. and Montaggioni, L. F. (1982) Crustal block movements from Holocene shorelines: Crete and Antikythira (Greece). Tectonophysics, **86**, 27–43.

Pitman, J. I. (1985) Thailand, in Bird and Schwartz (1985).

Polcyn, F. C. (1981) Applications of remote sensing techniques to coastal areas management, in S. Vallejo (ed.) *Coastal Management in Ecuador:* UN Ocean Economics and Technology Branch, New York.

Pomar, J. M. (1962) Cambios en los rios y en la morfologia de la costa de Chile, *Rev. Chil. Hist. y Geog.,* **130**, 318–56.

Prêcheur, C. (1960) Le littoral de la Manche: de Ste. Adresse à Ault. Etude morphologique. *Norios,* special volume.

Price, D. J. (1975) The apparent growth of Gulf Beach, extreme West Florida, *Trans. Gulf Coast Ass. Geol. Socs.,* **25**, 369–71.

Pringle, A. (1981) Beach development and coastal erosion in Holderness, North Humberside, in J. Neale and J. Flenley (eds.) *The Quaternary in Britain,* pp. 194–205.

Pringle, A. W. (1983) *Sand spit and bar development along the east Burdekin delta coast, Queensland.* James Cook University, Department of Geography, Monograph 13, Pergamon, Oxford.

Prior, D. B. (1975) Coastal landslide morphology and processes on Eocene clays in Denmark, *Geogr. Tidsskr.,* **76**, 14–33.

Psuty, N. P. (1965) Beach ridge development in Tabasco, Mexico, *Ann. Ass. Am. Geogr.* **55**, 112–24.

Pugh, J. C. (1954) A classificiation of the Nigerian coastline, *J. West African Sci. Assoc.,* **1**, 3–12.

Purser, B. H. and Evans, G. (1973) Regional sedimentation along the Trucial Coast, in B. H. Purser (ed.) *The Persian Gulf,* Springer Verlag, New York, pp. 211–31.

Quigley, R. M. and Di Nardo, L. R. (1980) Cyclic instability modes of eroding clay bluffs, Lake Erie Northshore Bluffs at Port Bruce, Ontario, Canada, *Z. Geomorph,* Supp. **34**, 39–47.

Raj. J. K. (1982) Net directions and rates of present-day beach transport by littoral drift along the East Coast of Peninsular Malaysia, *Bulletin, Geological Society of Malaysia,* **15**, 57–70.

Ranwell, D. S. (1974) The salt marsh to tidal woodland transition, *Hydrobiol. Bull.,* **8**, 139–51.

Rex, R. W. (1964) Arctic beaches, Barrow, Alaska, in R. L. Miller (ed.) *Papers in Marine Geology* Macmillan, New York, pp. 304–400.

Richards, A. F. (1960) Rates of marine erosion of tephra and lava at Isla San Benedicto, Mexico, *Int. Geol. Congr.* **21**, Norden, Subm. Geol., 58–63.

Richie, W. (1985) Scotland, in Bird and Schwartz (1985).

Rohde, H. (1978) The history of the German coastal regions, *Die Küste,* **32**, 6–29.

Rojdestvensky, A. B. (1972) Movements of coastal sands in the Gulf of Varna, *Proc. Inst. Oceangr. Fisheries, Varna,* **11**, 33–42 (in Bulgarian).

Ron, Z. (1982) Erosion of the beach cliff in Netanya and its regression, in A. Shmueli and M. Braver (eds.). *Netanya,* Am Oved, Tel Aviv, pp. 45–67.

Royal Commission on Coastal Erosion (1907–11) *Reports* (3 volumes), HMSO, London.

Rudberg, S. (1967) The cliff coast of Gotland and the rates of cliff retreat, *Geogr. Ann.,* **49A**, 283–98.

Russell, R. J. (1967) Aspects of coastal morphology, *Geogr. Ann.,* **49A**, 299–309.

Russell, R. J. (1970) *Oregon and Northern California coastal reconnaissance.* Coastal Studies Institute, Louisiana State University, Tech. Report. 86.

Sanlaville, P. (1982) Syria and Lebanon, in M. L. Schwartz (ed.) *Encyclopaedia of Beaches and Coastal Environments,* Hutchinson and Ross, Shroudsburg, Pennsylvania, pp. 98–102.

Sanlaville, P. (1985a) Arabian Gulf Coasts, in Bird and Schwartz (1985).

Sanlaville, P. (1985b) Syria and Lebanon, in Bird and Schwartz (1985).

Sambasiva Rao, M. and Vaidyanadhan, R. (1979) Morphology and evolution of the Godavari delta, India, *Z. Geomorph.*, **23**, 243–255.

Schnack, E. J. (1985) Argentina, in Bird and Schwartz (1985).

Schofield, J. C. (1967) Sand movement at Mangatawhiri Spit and Little Omaha Bay, *N.Z. J. Geol. Geophys.*, **10**, 697–721.

Schwartz, M. L. (1967) The Bruun theory of sea level rise as a cause of shore erosion, *J. Geol.*, **75**, 76–92.

Schwartz, M. L. (1971) Shannon Point cliff recession, *Shore and Beach*, **39**, 45–48.

Schwartz, M. L. (1985a) Pacific Colombia, in Bird and Schwartz (1985).

Schwartz, M. L. (1985b) Libya, in Bird and Schwartz (1985).

Schwartz, M. L. and Terich, T. A. (1985) Washington, in Bird and Schwartz (1985).

Seeling, W. N. and Sorenson, R. M. (1973) *Historic shoreline changes in Texas,* Texas A & M University.

Sherlock, R. L. (1922) *Man as a Geological Agent,* Witherby, London.

Shepard, F. P. and Wanless, H. R. (1971) *Our changing coastlines,* McGraw-Hill, New York.

Shinn, E. A. (1973) Sedimentary accretion along the leeward south-east coast of Qatar Peninsula, Persian Gulf, in B. H. Purser (ed.) *The Persian Gulf,* Springer Verlag, New York, pp. 199–211.

Short, A. D. (1979) Barrier-island development along the Alaskan-Yukon coastal plains. *Bull. Geol. Soc. America,* **90**, 77–103.

Shuijskii, J. and Simeonova, G. (1982) On the types of abrasion cliffs along the Bulgarian Black Sea Coast, *Bulg. Acad. Sci., Inj., Geol., Hidrogeol.*, **12**, 11–12

Simeonova, G. (1985) Bulgaria, in Bird and Schwartz (1985).

Sioli, H. (1966) General features of the delta of the Amazon, *Sci. Problems, Humid Tropical Zone Deltas,* pp. 381–90.

Snead, R. E. (1967) Recent morphological changes along the coast of West Pakistan, *Ann. Assoc. Amer. Geogr.*, **57**, 550–65.

Snead, R. E. (1969) *Physical Geography Reconnaisance: West Pakistan Coastal Zone.* Univ. New Mexico, Publications in Geography 1.

Snead, R. E. (1970) *Physical Geography of the Makran Coastal Plain of Iran.* U.S. Department of Commerce.

Snead, R. E. (1985) Bangladesh, in Bird and Schwartz (1985).

So, C. L. (1981) Coast changes between Lu Fau Shan and Nim Wan, Hong Kong. *Sains Malaysiana,* **10**, 51–68.

So, C. L. (1983) Beach changes and associated problems of beach conservation in Hong Kong. *Proc. 7th Annual Conference, Coastal Society of America,* pp. 127–131.

Stafford, D. E. (1972) *A state of the art survey of the applications of aerial remote sensing to coastal engineering.* Coastal Engineering Research Center, Fort Belvoir, U.S.A.

Stapor, F. (1971) Sediment budgets on a compartmented low-to-moderate energy coast in northwest Florida, *Marine Geol.*, **10**, M1–M7.

Stapor, F. (1975) Shoreline changes between Phillips Inlet and Pensacola Inlet, Northwest Florida coast, *Trans. Gulf Coast Ass. Geol. Socs.*, **25**, 373–8.

Steers, J. A. (ed.) (1960) *Scolt Head Island,* Heffer, Cambridge.

Steers, J. A. (1964) *The coastline of England and Wales,* Cambridge University Press, Cambridge.

Steers, J. A. (1973) *The coastline of Scotland,* Cambridge, University Press, Cambridge.

Steers, J. A., Stoddart, D. R., Bayliss-Smith, T. P., Spencer, T. and Durbridge, P. H. (1979) The storm surge of 11 January 1978 on the east coast of England, *Geogr. J.*, **145**, 197–205.

190

Stembridge, J. E. (1975) Recent shoreline changes of the Alsea sandspit, Lincoln County Oregon, *The Ore Bin*, **37**, 77–82.
Stembridge, J. E. (1976) *Recent shoreline changes of the Oregon coast*, U.S.A. East Carolina University.
Stoddart, D. R. (1964) Storm conditions and vegetation in equilibrium of reef islands. *Proc. 9th Conference Coastal Engineering*, pp. 893–906.
Stoddart, D. R., McLean, R. F., Scoffin, T., Thom, B. G., and Hopley, D. (1978) Evolution of reefs and islands, northern Great Barrier Reef, *Phil. Trans. Roy. Soc. Lond., B*, **291**, 149–60.
Sunamura, T. and Horikawa, K. (1977) Sediment budget in Kujukuri coastal area, Japan, *A.S.C.E. Coastal Sediments '77*, pp. 475–87.
Swan, S. B. (1965) Coast erosion principles and classification of south-west Ceylon's beaches on the basis of their erosional stability, *Ceylon Geogr.*, **19**, 1–16.
Swan, S. B. (1968) Coastal classification with reference to the east coast of Malaya, *Z. Geomorph., Supp.* **7**, 114–32.
Swan, S. B. (1971) Coastal morphology in a humid tropical low energy environment: the islands of Singapore, *J. Trop. Geogr.*, **33**, 43–61.
Swan, S. B. (1974) *The coast erosion hazard, southwest Sri Lanka*, University of New England Research Series in Applied Geography, pp. 40.
Swan, S. B. (1982) *The coastal geomorphology of Sri Lanka: an introductory survey*. University of New England, Armidale, Australia.

Tanaka, N. (1973–4) Investigations of shoreline change on the basis of aerial photographs, *Port & Harb. Research Inst. Japan, Tech. Notes*, **163**, 192.
Tanner, W. F. (1975) Symposium on beach erosion in Middle America, *Trans. Gulf Coast Ass. Geol. Socs.*, **25**, 365–8.
Tanner, W. F. (ed.) (1978) *Standards for Measuring Shoreline Changes*. Department of Geology, Florida State University, Tallahassee.
Tanner, W. F. and Stapor, F. (1971) Tabasco beach ridge plain: an eroding coast, *Trans. Gulf Coast Ass. Geol. Socs.*, **21**, 231–2.
Tanner, W. F. and Stapor, F. (1972) Accelerating crisis in beach erosion, *Internat. Geogr.*, **2**, 1020–1.
Teh Tiong Sa (1985) Malaysia, in Bird and Schwartz (1985).
Thom, B. G. (1968) Coastal erosion in eastern Australia, *Australian Geogr. Studies*, **6**, 171–3.
Thom, B. G. (1969) Problems of the development of Isla del Carmen, Campeche, Mexico, *Z. Geomorph.*, **13**, 406–13.
Thom, B. G. (1973) Contemporary coastal erosion—geologic or historic? *Impact of Human Activities on Coastal Zones* (UNESCO/MAB) pp. 58–65.
Thom, B. G. (1974) Coastal erosion in eastern Australia, *Search*, **5**, 198–29.
Thom, B. G. (1978) Coastal sand deposition in south-east Australia during the Holocene, in J. L. Davies and M. A. J. Williams, *Landform Evolution in Australasia*, pp. 197–214.
Thom, B. G. and Wright, L. D. (1983) Geomorphology of the Purari delta, in T. Petr (ed.) *The Purari—Tropical Environment of a High Rainfall River Basin*, Dr Junk, The Hague, pp 47–65.
Thomas A. (1983) *Contribution a une étude d'impact des activites humaines au littoral du Cap de l'Aigle (Bouches-du-Rhône) au Cap Garonne (Var): la dynamique sédimentaire de l'herbier a Posidonies*. Thèse Univ. Aix-Marseille.
Tinley, K. (1985). Mocambique, in Bird and Schwartz (1985).
Tjia, H. D. (1973) Geomorphology, in D. J. Gobbert and C. S. Hutchinson, (eds.) *Geology of the Malay Peninsula*, Wiley-Interscience, New York, pp. 13–24.
Troll, C. and Schmidt-Kraepelin, E. (1965) Das neue delta des Rio Sinu an der Karibischen Küste Kolumbiens, *Erdkunde*, **19**, 14–23.

Tziavos, C. (1977) *Sedimentology, ecology, and palaeogeography of the Sperchios Valley and Maliakos Gulf, Greece*. Univ. Delaware, U.S.A.

U.S. Corps of Engineers (1971) *National Shoreline Survey*, Washington, DC.

Usoro, E. J. (1985) Nigeria, in Bird and Schwartz (1985).

Valentin, H. (1952) *Die Küsten der Erde*, Petermanns Geogr. Mitt., p. 246.

Valentin, H. (1953) Present vertical movements of the British Isles, *Geogr., J.*, **119**, 299–305.

Valentin, H. (1954) Die landverlust in Holderness, Ostengland, von 1852 bis 1952, *Die Erde*, **3**, 297–315.

Vanney, J. R., Menenteau, L. and Zazo, C. (1979) Physiographie et évolution des dunes de base Andalousie (Golfe de Cadix), in A. Guilcher (ed.) *Les Côtes Atlantiques de l'Europe*, pp. 277–86.

Van Straaten, L. M. J. U. (1959) Littoral submarine morphology of the Rhône delta, *2nd Coastal Geography Conference, Proceedings*, pp. 233–64.

Verger, F. (1968) *Les marais des côtes francaises de l'Atlantique et de la Manche et leurs marges maritimes*, Thèse d'Etat, Paris, Bordeaux.

Verstappen, H. T. (1953) *Djakarta Bay: a geomorphological study on shoreline development*, Rijkuniversitei, Utrecht.

Verstappen, H. T. (1966) The use of aerial photographs in delta studies, *Sci. Problems Humid Tropical Zone Deltas*, pp. 373–9.

Verstappen, H. T. (1977) *Remote sensing in geomorphology*. Elsevier, Amsterdam.

Verstappen, H. T. (1973) A geomorphological reconnaissance of Sumatra and adjacent islands, *Verh. K. Ned. Aardijk. Genoot.*, 1.

Volker, A. (1966) The deltaic area of the Irrawaddy River in Burma, *Sci. Problems Humid Tropical Zone Deltas*, pp. 373–9.

Voss. F. (1970) The influence of the latest transgression phase on the coastal evolution of the Geltinger Birck, in the northern part of the western Baltic, *Die Küste*, **20**, 101–13.

Wadsworth, A. H. (1966) Historical deltation of the Colorado River, Texas, in M. L. Shirley (ed.) *Deltas in their Geologic Framework*, Houston Geological Society, Texas, pp. 99–105.

Wagle, B. G. (1982) Geomorphology of the Goa coast, *Proc. Indian Acad. Sci.*, **91**, 105–17.

Walker, H. J. (1981) Man and shoreline modification, in E. C. F. Bird and K. Koike (eds.). *Coastal Dynamics and Scientific Sites*, Komazawa University, Tokyo, pp. 55–90.

Walker, H. J. (1985) Alaska, in Bird and Schwartz (1985).

Walker, H. J. and McCloy, J. M. (1969) *Morphologic change on two arctic deltas*, Arctic Inst. of N. America, Research Paper **49**.

Webb, J. E. (1960) *The erosion of Victoria Beach: its causes and cure*, Ibadan, University Press, Nigeria.

Weichset, W. (1963) Further observations on geologic and geomorphic changes from the catastrophic earthquake of May 1960 in Chile, *Bull. Seism. Soc. America*, **53**, 1237–57.

Weisrock, A. L. E. (1985) Morocco, in Bird and Schwartz (1985).

Weisser, P. J., Garland, I. F. and Drews, B. K. (1982) Dune advancement 1937–1977 at the Mlalazi Nature Reserve, Mtunzini, Natal, *Bothalia*, **14**, 127–30.

Wells, J. and Coleman, J. (1981) Periodic mudflat progradation, northeastern coast of South America: a hypothesis, *J. Sed. Petrol.*, **51**, 1053–75.

West, R. C. (1956) Mangrove swamps of the Pacific coast of Colombia, *Ann. American Assoc. Geogr.*, **46**, 98–121.

Wilkinson, B. H. and McGowen, J. H. (1977) Geologic approaches to the determination of long-term coastal recession rates, Matagorda Peninsula, Texas, *Environ, Geol.*, **1**, 359–65.

Williams, A. T. and Davies, P. (1980) Man as a geological agent: the sea cliffs at Llantwit Major, Wales, *Z. Geomorph.*, Supp. **34**, 129–41.

Williams, P. W. (1977) Progradation of Whatipu Beach, Manukau Harbour, *New Zealand J. Marine Freshw. Research*, **12**, 429–50.

Wiseman, G., Hayati, G. and Frydman, S. (1981) Stability of a heterogeneous sandy coastal cliff, *Proc. 10th Int. Conf. Soil Mechanics*, pp. 569–74.

Wong, Poh Poh (1981) Beach changes on a monsoon coast, Peninsular Malaysia, *Bulletin, Geological Society of Malaysia*, **14**, 59–74.

Wong, Poh Poh (1985) Singapore, in Bird and Schwartz (1985).

Wood, P. A. (1976) Beaches of accretion and progradation in Jamaica, *J. Geol. Soc. Jamaica*, **15**, 24–31.

Wright, L. W. (1969) Coastal changes at the entrance to the Kaipara Harbour, 1836–1966, *New Zealand Geogr.*, **25**, 58–61.

Yamanouchi, H. (1977) A geomorphological study about the coastal cliff retreat along the south west coast of the Atsumi Peninsula, Central Japan, *Sci. Rep. Fac.Educ. Gumma University*, **26**, 96–128.

Zaneveld, J. S. and Verstappen, H. T. (1952) A recent investigation on the geomorphology and the flora of some islands in the Bay of Jakarta, *J. Sci. Research*, **3**, 58–68.

Zeigler, J. M., Tuttle, S. D., Giese, G. S. and Tasha, H. J. (1964) Residence time of sand composing beaches and bars of Outer Cape Cod, *Proc. 9th Conf. Coast, Eng.*, **26**, 403–16.

Zenkovich, V. P. (1958) *The Black and Azov Sea shores*, Geografgiz, Moscow.

Zenkovich, V. P. (1967) *Processes of Coastal Development* (trans. O. G. Fry; ed. J. A. Steers), Edinburgh.

Zenkovich, V. P. (1973) Geomorphological problems of protecting the Caucasian Black Sea coast, *Geogr. J.*, **139**, 460–6.

Zenkovich, V. P. (1976) Preserving the nature of seashores, *Geoforum*, **7**, 395–7.

Zenkovich, V. P. (1985) Eastern Black Sea, U.S.S.R., in Bird and Schwartz (1985).

Zunica, M. (1971) *La spiagge del Venete*, Presso Istituto de Geografia, Università di Padova.

Zunica, M., (1976) Coastal changes in Italy during the past century, *Italian Contributions, 23rd I.G.C.*, pp. 275–281.

Appendix

On behalf of the IGU Commission on the Coastal Environment, the author would like to acknowledge contributions from the following correspondents:

AHMAD, E. Ranchi University (India)
ALEXANDER, C. S. University of Illinois, Urbana (Tanzania)
AL KAISI, K. Arabian Gulf University (Arabia)
ANDREW, W. Public Works Department, Perth (W. Australia)
ARAYA-VERGARA, J. F. University of Santiago (Chile)
ÅSE, L. E. University of Uppsala, Sweden (Sweden)
ASENSIO AMOR, I. University of Madrid (Spain)
ATLANTIDA, C. de HURTADO, National University, Mexico City (Mexico)
BATTISTINI, R. University of Orleans (Madagascar)
BIRD, J. B. McGill University, Montreal (Arctic Canada)
BODERE, J. C. University of Western Brittany (Iceland)
BORÓWCA, R. K. Adam Mickiewiez University (Poland)
BOTTS, L. Lake Michigan Federation, Chicago (Great Lakes)
BREMNER, J. M. University of Cape Town (South Africa)
BRESSOLIER, C. Geomorphology Laboratory, Dinard (France)
BRUUN, P. Technical University of Norway (Norway)
BRYANT, E. University of Wollongong, Australia (Canada)
BUCKLES, W. R. Bowling Green State University, Ohio (Great Lakes)
CAMPBELL, J. F. University of Hawaii (Hawaii)
CARTER, C. H. Ohio Geological Survey (Great Lakes)
CARTER, R. W. G. New University of Ulster (Ireland)
CHEN JIYU, East China Normal University, Shanghai (China)
CRAIG, A. Florida Atlantic University (Peru; Bahamas)
CRUZ, O. University of São Paulo (Brazil)
DAGORNE, A. University of Nice (Algeria)
DAVIES, J. L. Macquarie University, Sydney (Tasmania)
DEI, L. A. University of Cape Coast (Ghana)
DIONNE, J. C. Environment Canada (E. Canada)
DOLAN, R. Environmental Sciences, Charlottesville, Virginia (Atlantic U.S.A.)
DUBOIS, J. M. M. University of Sherbrooke, Quebec (E. Canada)
ELIOT, I. G. University of Western Australia (W. Australia)
EROL, O. University of Ankara (Turkey)
EVERTS, C. Coastal Engineering Research Center (Atlantic U.S.A.)
FABBRI, P. University of Bologna (Italy)
FISHER, J. J. University of Rhode Island (Atlantic U.S.A.)
GALLOWAY, R. W. C.S.I.R.O., Canberra (Australia)
GIERLOFF-EMDEN, H. G. University of Munich (Central America)
GILLIE, R. D. McMaster University, Canada (Great Lakes)
GOLDSMITH, V. Israel Oceanographic Research, Haifa (Atlantic U.S.A., Israel)
GRANÖ, O. University of Turku (Finland)
GUILCHER, A. University of W. Brittany (France; W. Africa; Tahiti)
HARVEY, W. Flinders University, Bedford Park (S. Australia)
HEALY, T. R. University of Waikato (New Zealand)
HEYDORN, A. E. F. Institute for Oceanology, Stellenbosch (South Africa)

HINSCHBERGER, F. University of Caen (Ivory Coast)
HOPLEY, D. James Cook University, Townsville (N. Australia)
ILTIS, J. ORSTOM, Noumea (New Caledonia)
JENNINGS, J. N. Australia National University (Australia)
KAPLIN, P. Moscow State University (U.S.S.R.: Pacific Islands)
KING, C. A. M. University of Nottingham (Iceland)
KIRK, W. University of Canterbury, New Zealand (Antarctic)
KLEMSDAL, T. University of Oslo (Norway)
KOIKE, K. Komazawa University, Tokyo (Japan)
KRAFT, J. C. University of Delaware (Greece; Turkey)
KOMAR, P. D. Oregon State University (Oregon)
LARSEN, C. E. University of North Carolina (Great Lakes)
LEONTIEV, O. K. Moscow State University (Caspian)
LEATHERMAN, S. P. University of Maryland (Atlantic U.S.A.)
LÖFFLER, E. C.S.I.R.O., Canberra (Papua New Guinea)
MARQUES, M. A. University of Barcelona (Spain)
MAY, V. J. Bournemouth College of Technology (British Isles)
McCANN, S. B. McMaster University (Canada)
McHONE, J. National Oceanic & Atmospheric Administration, Florida (Virginia)
McLEAN, R. F. University of Auckland (New Zealand)
MISTARDIS, G. G. Institute of Oceanography, Athens (Greece)
MØLLER, J. T. Aarhus University (Denmark)
MOLNIA, B. F. U.S. Geological Survey, Menlo Park (Alaska)
MROCZEK, P. University of Berlin (Germany)
MUEHE, D. City University, Rio de Janeiro (Brazil)
NORRMAN, J. O. University of Uppsala (Sweden, Iceland)
NOSSIN, J. J. International Institute for Aerial Survey (Netherlands)
OGDEN, J. G. Dalhousie University, Canada (Atlantic U.S.A.)
OJANY, F. J. University of Nairobi (Kenya)
ONGKOSONGO, O. S. R. Oceanographic Institute, Jakarta (Indonesia)
ORME, A. R. University of California, Los Angeles, U.S.A. (California)
OWENS, E. H. Bedford Institute of Oceanography (Canada)
OZER, A. University of Liège (Belgium)
PARK, DONG WON Seoul National University (Korea)
PASKOFF, R. University of Tunis (Mediterranean; Chile)
PIRAZZOLI, P. Geomorphology Laboratory, Montrouge (Mediterranean)
POMEL, R. University of Abidjan (Ivory Coast)
PSUTY, N. P. Rutgers University, New Jersey (Latin America)
PUGH, J. C. University of London (Nigeria)
RAJ. U. Marine Science Institute, Suva (Fiji)
REDFIELD, A. C. Woods Hole Oceanographic Institution (Venezuela)
RITCHIE, W. University of Aberdeen (Scotland)
ROJDESTVENSKY, A. V. Institute of Oceanography, Varna (Bulgaria)
SCHNACK, E. Centre for Coastal Geology, Mar del Plata (Argentina)
SCHWARTZ, M. L. Western Washington University (Pacific U.S.A.)
SEGADO, M. Instituto Hidrografico de la Marina, Cádiz (Spain)
SHORT, A. D. Macquarie University, Sydney, Australia (Alaska)
SHUISKY, Y. D. Odessa University (U.S.S.R.: Albania)
SIMEONOVA, G. Academy of Sciences, Sofia (Bulgaria)
SNEAD, R. E. University of New Mexico (Pakistan, Bangladesh)
SO, C. L. University of Hong Kong (Hong Kong)
SOEGIARTO, A. Oceanographic Institute, Jakarta (Indonesia)
STAPOR, F. Florida State University (Florida)
STEMBRIDGE, J. E. East Carolina University (Oregon)

STEPHENS, N. University College, Swansea (Iceland)
SUNAMURA, T. Institute of Geoscience, Tsukuba (Japan)
SWAN, S. B. University of New England, Armidale (Sri Lanka; Malaysia)
TANNER, W. F. Florida State University (Mexico; Caribbean)
TEH TIONG SA. University of Malaysia, Kuala Lumpur (Malaysia)
THOM, B. G. Royal Military College, Canberra (Australia, Papua New Guinea)
TWIDALE, C. R. University of Adelaide (South Australia)
TZIAVOS, C. Institute of Oceanography, Athens (Greece)
URIEN, C. M. Buenos Aires (Argentina)
VAIDYANADHAN, R. Andhra University, Waltair (India)
VERGER, F. Geomorphology Laboratory, Dinard (France)
WAGLE, B. G. Institute of Oceanography, Goa, (India)
WALKER, H. J. Louisiana State University, Baton Rouge (Alaska)
WEISROCK, A. University of Lyon (Morocco)
WONG POH POH University of Singapore (Singapore, Malaysia)
YAMANOUCHI, H. Gumma University (Japan)
YASSO, W. E. Columbia University (Atlantic U.S.A.)
ZEIGLER, J. M. Virginia Institute of Marine Science (Atlantic U.S.A.)
ZENKOVICH, V. P. Academy of Sciences, Moscow (U.S.S.R.; China)
ZUNICA, M. University of Padua (Italy)

Author Index

Ahlmann, H. W., 44
Ahmad, E., 110
Alestalo, J., 48
Alexander, C. S., 36, 107
Al Kaisi, K., 193
Andrew, W., 137, 193
Andrews, J. T., 42
Araya-Vergara, J. F., 27, 28, 29
Armon, J. W., 41, 150
Armon, J. W., and McCann, S. B., 41
Åtse, L. E., 46, 47
Asensio Amor, I., 78, 193
Atlantida, C. de Hurtado, 193
Ayon, H., and Jara, W., 27

Baines, G. B. K., and McLean, R. F., 154, 155
Ballard, P., 140
Baltzer, F., 151
Banks, R. S., 36
Battistini, R., and Bergoeing, J. P., 26
Battistini, R., and Le Bourdiec, P., 107
Bedi, N., and Vaidyanadhan, R., 111
Beke, C. T., 109
Berry, R. W., et al., 109
Bird, E. C. F., 1, 3, 11, 50, 125, 132, 140, 147, 169
Bird, E. C. F., and Barson, M., 140
Bird, E. C. F., and Christiansen, C., 52
Bird, E. C. F., Dubois, J. P., and Iltis, J. A., 151
Bird, E. C. F., and Guilcher, A., 107
Bird, E. C. F., and May, V. J., 10, 59
Bird, E. C. F., and Ongkosongo, O. S. R., 126, 129
Bird, E. C. F., and Paskoff, R., 160, 169
Bird, E. C. F., and Ramos, V. T., 27
Bird, E. C. F., and Rosengren, N., 128
Bird, E. C. F., and Schwartz, M. L., 11

Bird, J. B., Richards, A., and Wong, P. P., 37
Bird, Juliet, 70
Blanc, J. J., and Grissac, A. J., 79
Bodéré, J. C., 44
Borówca, R. K., 50
Botts, L., 193
Bourman, R. P., 140
Bremner, J. M., 103
Bressolier, C., 193
Briquet, A., 76
Broggi, J. A., 27
Brown, M. J. F., 133, 134
Brunsden, D., and Jones, D. K. C., 64, 66, 67
Bruun, P., 43, 46, 169
Bryan, R. B., and Price, A. G., 43
Bryant, E. A., 41, 145, 173
Buckles, W. R., 43, 193
Byrne, J. V., 20

Caldwell, J. M., 38
Calles, B., Linde, K., and Norrman, J. O., 45
Cambers, G., 63
Cameron, H. L., 156
Campbell, J. F., 153
Campbell, J. F., and Hwang, D. J., 153
Caputo, C., et al., 81
Carr, A. P., 6, 68
Carr, A. P., and Gleason, R., 72
Carter, C. H., 42
Carter, R. W. G., 68, 74
Carter, R. W. G., et al., 76
Cencini, C., et al., 83
Chamberlain, T., 153
Chambers, M. J. G., and Sobur, A. J., 126
Chapman, D. M., et al., 145
Chappell, J., 132
Chen Jiyu, et al., 118

Christiansen, C., *et al.*, 55
Christiansen, C., and Møller, J. T., 52
Clayton, K. M., 70, 71
Coakley, J. P., 42
Craig, A. K., 37
Craig, A. K., and Psuty, N. P., 27
Cruz, O., *et al.*, 30, 31

Daetwyler, C. C., 20
Davies, J. L., 143
Deane, C. A. W., 36
De Boer, G., 62
Dei, L. A., 102
Del Canto, S., and Paskoff, R., 28
Depuydt, F., 59
Dionne, J. C., 41
Diresse, P., and Kouyoumontzakis, G., 103
Dolan, R., and Bosserman, K., 4
Dolan, R., *et al.*, 38
Dresch, J., 27
Dubois, J. M., 41, 169
Dunbar, G. S., 38

Edelman, T., 10, 57, 58
Eisma, D., 97, 117
Eisma, D., and Park, D. W., 119
El Ashry, M. T., 7
Eliot, I. G., *et al.*, 136
Ellenberg, O., 31
Emery, K. O., 24
Emery, K. O., and Neev, D., 97
Erol, O., 95, 96, 97
Evans, G., 97
Everard, C. E., 72
Everts, C., 39, 193

Fabbri, P., 81, 193
Facon, R., 78
Fairbridge, R. W., 169
Fels, E., 84
Fisher, J. J., 37
Fisher, J. J., and Regan, D. R., 7
Fisher, J. J., and Simpson, E. J., 40
Froomer, N., 38, 174

Galloway, R., 134, 193
Gastescu, P., and Breier, A., 88
Gibb, J. G., 147, 149
Gierloff-Emden, H. G., 26, 50, 57
Gilbert, G. K., 21
Gillie, R. D., 42, 193
Goldsmith, V., 98
Goldsmith, V., and Golik, A., 97–98

Goldsmith, V., *et al.*, 38
Granö, O., 49
Griggs, J. B., and Johnson, R. E., 21
Gudelis, Y., 49, 50
Guilcher, A., 59, 77, 78, 79, 104, 108, 175
Guilcher, A., and Nicolas, J. P., 101
Guilcher, A., *et al.*, 103
Gulliver, F. P., 8

Hallegouët, B., and Moign, A., 77
Harvey, N., 140
Hayes, M. O., 37
Healy, T. R., 147
Heerden, I. L. V., and Roberts, H. H., 35
Helle, R., 48
Hehanussa, P. E., 129, 130
Herd, D. G., *et al.*, 26
Hernandez Pacheca, F., 78
Heusser, C., 29
Heydorn, A. E. F., and Tinley, K. L., 105
Hicks, S. D., and Shofnos, W., 17
Hinschberger, F., 101
Hodgkin, E., 137
Höllergwoger, F., 130
Hsu, T. L., 119
Hunter, J. F., 38
Hutchinson, J. N., 63, 64
Hwang, D., 153

Iriondo, M., and Scotta, E., 29

Jackson, J. M., 29
Jennings, J. N., 4, 135, 137
Johnson, D. W., 6, 8
Johnson, J. W., 20
Johnston, W. A., 18
Jones, M., 47, 48, 49

Kaplin, P., 124, 154
Kaye, C. A., 40
Kelletat, D., 87
Kestner, F. J. T., 176
Khafagy, A., and Manohar, M., 98
King, C. A. M., 42, 44
Kirk, R. M., 150
Klemsdal, T., 46
Koike, K., 119, 120, 122
Komar, P. D., 19
Koopmans, B. N., 113, 115
Kraft, J. C., 37
Kraft, J. C., *et al.*, 84, 85, 95

Kramer, J., 57
Kukla, G., and Gavin, J., 157

La Fond, E. C., 112
Larsen, C., 42, 43
Leatherman, S. P., 194
Le Bourdiec, P., 101
Lees, G. M., and Falcon, N. L., 109
Leontiev, O. K., 92, 93, 94
Leontiev, O. K., et al., 91
Lewellen, R., 13
Lindh, G., 46
Löffler, E., 132
Luck, G., 52
Lustig, T. L., 135
Luternauer, J. L., 18

MacCarthy, G. R., 13, 14
Macdonald, H. V., et al., 145
Mackay, J. R., 42
Mahrour, M., and Dagorne, A., 100
Mamykina, V. A., 91
Manley, G., 173
Marker, M. E., 74
Marques, M. A., and Julia, R., 79
Marshall, P., 147
Martin, L., et al., 31
Martini, I. P., 41
Mathlouti, S., and Paskoff, R., 100
May, S. K., et al., 10
May, V. J., 60, 64, 176
Mazzanti, R., and Pasquinucci, M., 81
McCann, S. B., 41
McCann, S. B., and Hale, P. B., 18
McCormick, C. L., 38
McHone, J., 38, 194
McIntire, W. G., and Walker, H. J., 156
McLean, R. F., 147, 150, 151, 152
Meijerink, A. M. J., 111
Meinesz, A., and Lefevre, J. R., 79
Meinesz, A., et al., 79
Miller, D. J., 16
Miossec, J., and Paskoff, R., 100
Mistardis, G. G., 194
Moberly, R., and Chamberlain, T., 153
Moign, A., and Guilcher, A., 46
Moldovanu, A., and Selariu, O., 88
Molnia, B. F., 15, 16, 17
Møller, J. T., 52, 56
Morelock J., 36
Morelock, J., and Trumbull, J. V. A., 36
Morgan, J. P., 33, 34
Morton, R. A., 33

Mroczek, P., 57
Muehe, D., 30

Nagaraja, V. N., 111
Nageswara Rao, K., and Vaidyanadhan, R., 111
Nielsen, E., 98
Nielsen, N., 44
Nir, Y., 98
Nordstrom, K. F., and Allen, J. R., 38
Norris, R. M., 22
Norrman, J. O., 44, 45, 46
Nossin, J. J., 115
Nummedal, D., and Stephen, M. F., 15

Oertel, G. F., and Chamberlain, C. F., 37
Ogden, J. G., 40
Ojany, F. J., 107, 194
Olson, J. S., 43
Ongkosongo, O. S. R., 132
Orford, J. D., and Carter, R. W. G., 76
Orlova, G., and Zenkovich, V. P., 98
Orme, A. R., 20, 24, 106, 166
Ottmann, F., 4, 76
Owens, E. H., 41
Owens, E. H., and Bowen, A. J., 41
Owens, E. H., and Harper, J. R., 18
Ozasa, H., 120
Ozer, A., 59

Park, D. W., 119
Parker, R., 64
Paskoff, R., 100
Paskoff, R., and Sanlaville, P., 83
Petersen, M., 57
Pichler, H., et al., 87
Pierce, J. W., 38
Piper, D. J. W., and Panagos, A. G., 87
Pirazzoli, P. A., et al., 87
Pitman, J. I., 116
Polcyn, F. C., 8, 112
Pomar, J. M., 28
Pomel, R., 194
Prêcheur, C., 76
Price, D. J., 35
Pringle, A., 62, 146
Prior, D. B., 52
Psuty, N. P., 32
Pugh, J. C., 102
Purser, B. H., and Evans, G., 109

Quigley, R. M., and Di Nardo, L. R., 43

Raj, J. K., 114
Raj, U., 194
Ranwell, D. S., 74, 75, 174
Redfield, A. C., 194
Rex, R. W., 14
Richards, A. F., 25
Ritchie, W., 72, 194
Rohde, H., 57
Rojdestvensky, A. B., 87
Ron, Z., 98
Rosengren, N., 128, 129, 143
Rudberg, S., 46
Russell, R. J., 19, 173

Sambasiva Rao, M., and Vaidyanadhan, R., 111
Sanlaville, P., 97, 110
Schnack, E. J., 29
Schofield, J. C., 147
Schwartz, M. L., 11, 18, 26, 43, 99, 169
Schwartz, M. L., and Terich, T., 19
Seelig, W. N., and Sorensen, R. M., 189
Segado, M., 78
Shepard, F. P., and Wanless, H. R., 10, 14, 18, 20, 22, 33, 35, 38, 40, 153
Sherlock, R. L., 9
Shinn, E. A., 109
Short A. D., 13
Shuijskii, J., and Simeonova, G., 87
Shuisky, Y. D., 84, 89, 90, 91, 124
Simeonova, G., 88
Sioli, H., 31
Snead, R. E., 110, 112
So, C. L., 117, 118
Stafford, D. E., 7
Stapor, F., 35
Steers, J. A., 10, 59, 63, 68, 72
Steers, J. A., et al., 68
Stembridge, J. E., 19, 20
Stephens, N., 74, 194
Stoddart, D. R., 32
Stoddart, D. R., et al., 146
Sunamura, T., and Horikawa, K., 121
Swan, S. B., 111, 112, 114, 115, 116

Tanaka, N., 122
Tanner, W. F., 25, 31, 32, 36, 37

Tanner, W. F., and Stapor, F., 30, 32, 35
Teh Tiong Sa, 115, 116
Thom, B. G., 32, 143, 147, 173
Thom, B. G., and Wright, L. D., 134
Thomas, A., 79
Tinley, K., 106
Tjia, H. D., 113
Troll, C., and Schmidt-Krapelin, E., 31
Twidale, C. R., 195
Tziavos, C., 54, 56

Urien, C. M., 195
Usoro, E. J., 102

Valentin, H., 1, 60, 61, 72
Vanney, J. R., et al., 78
Van Straaten, L. M. J. U., 79
Verger, F., 77
Verstappen, H. T., 126, 129, 130, 162
Volker, A., 113
Voss, F., 50

Wadsworth, A. H., 33
Wagle, B. G., 111
Walker, H. J., 13, 15, 175
Walker, H. J., and McCloy, J. M., 14
Webb, J. E., 102
Weichset, W., 28
Weisrock, A. L. E., 101
Weisser, P. J., et al., 106
Wells, J., and Coleman, J., 31
West, R. C., 26
Wilkinson, B. H., and McGowen, J. H., 33
Williams, A. T., and Davies, P., 68
Williams, P. W., 147
Wong Poh Poh, 114, 115
Wood, P. A., 36
Wright, L. W., 147, 149

Yamanouchi, H., 122
Yasso, W. E., 195

Zaneveld, J. S., and Verstappen, H. T., 130
Zeigler, J. M., et al., 40
Zenkovich, V. P., 49, 50, 59, 91, 92, 117, 118, 123, 124, 159, 164
Zunica, M., 80, 81

Location Index

Abe R., 121
Abidjan, 101, 161
Abu Dhabi, 109
Abukuma, 120
Abulat Knoll, 108
Acapulco, 25
Acheloos R., 84, 87
Ada, 102
Adelaide, 140
Aden, 108
Aden, Gulf of, 108
Adige Delta, 80
Adji Lagoon, 92
Adour R., 78
Adria, 80
Adriatic Coast, 81
Agano R., 122
Agde, 79
Agigea, 88
Agrakhan Spit, 92
Agulhas, 105
Agusan R., 126
Ahururi Lagoon, 147
Aitape, 132
Akkar, Bay of, 97
Aklan delta, 126
Akzybir Lagoon, 92
Alaid I., 124
Åland, 47
Alaska, Gulf of, 15, 159
Alaska Peninsula, 14
Aldabra, I., 156
Aleutian Is., 14, 159
Algarve Coast, 78
Algiers, 100
Al Manamah, 109
Alsea Sand Spit, 20
Alvarado, 32
Amapa Province, 31
Amazon R., 31
Amirantes I., 156

Amvrahia, Gulf of, 84
Anak Krakatau, 127, 128
Anapa, 91
Anchorage, 14, 160
Andalusian Bight, 167
Angola, 103
Anholt I., 51
Año Neueo, Pt., 21
Anse de l'Aiguillon, 77
Antalya, 97
Antarctica, 159
Antofagasta, 28
Antwerp, 59
Anzio, 81, 82
Apalachee Bay, 33
Apalachicola, 36
Apam, 102, 166
Apollo Bay, 140
Aquilea, 80
Arabian Gulf, 168
Aran Is., 74
Aransas Pass, 33
Arena Pt., 20
Argonaut Plain, 18
Arica R., 28, 166
Aringay R., 125
Arco Muto, 81
Arnhem Land, 134
Arno delta, 81
Arno R., 81
Arroyo Grande R., 21
Ascension I., 154
Ashdod, 98
Ashqelon, 98
Aspronisi, 15, 87
Assumption I., 156
Astove I., 156
Aswan High Dam, 99, 163
Atacama Desert, 27
Atami, 123
Atchafalaya R., 35

Atsumi Peninsula, 122
Auckland Peninsula, 147
Auma Spit, 134
Ault, 76
Authie Estuary, 76
Avila Beach, 21
Awatere R., 149
Axarfjördur, 44
Axiós delta, 84
Azerbaiian, 92
Azores, 154
Azov, Sea of, 89, 90, 91, 168

Bafa, L., 95
Baffin Bay, 42
Bagana, Mt., 134
Bago R., 126
Bahama Banks, 168
Bahamas, 37
Bahia Blanca, 29
Bahia de Caráquez, 27
Bahia de Paracas, Peru, 27, 165
Bahia Grande, 29
Bahrein I., 109
Baia dos Tigres (Tiger Bay), 103, 104
Baja California, 25
Balasore, 111
Balinas Pt., 21
Bandar Abbas, 109, 110
Bandar-e-Lengeh, 109
Bandar-e-Shah, 92
Bandon, 20
Bankaderes delta, 130
Bangkok, 116
Bangsanga R., 126
Banks Peninsula, 150
Baratana Bay, 35
Barbados, 37
Barents Sea, 124
Barranguilla, 31
Barron delta, 146, 147
Barron R., 145
Bartin, 97, 160
Barton Bay, 64
Bassin d'Arcachon, 78
Bata, 103
Bataan Peninsula, 125
Batang Hari delta, 126
Bathurst I., 134
Batina, 109
Batrun, 97
Batumi, 91
Bauang R., 125
Bayocean Spit, 20

Beach of Passionate Love (Pantai Cinta Berahi) 115
Beaufort Sea, 13, 42
Beirut, 97
Belle Passe, 35
Bellingham Bay, 18
Benacre Ness, 68, 166, 167
Bengal, Bay of, 166
Benghazi, 99
Berbera, 108
Berck Plage, 76
Bergrivier, 105
Bermuda, 156
Bernado Fiord, 29
Beruwala, 112
Bibione, 80
Bideford Bay, 68
Biferno R., 81
Big Flat Ck., 20
Big Sur coast, 21
Biscay, Bay of, 78
Bizerta, 99
Blackwood R., 137
Black Sea, 87–91, 95
Blakeney, 69
Blakeney Pt., 68, 71, 72
Blind Pass, 36
Bodri, 130
Bogoslof I., 14, 161
Bohai, Gulf of, 118
Bonifacio, 80
Bonne Anse, 78
Bora Bora, 154
Borra Principal, 32
Bosok, 130
Boston Harbour, 40
Bothnia, Gulf of, 46, 47, 160
Bouameu R., 151
Boubjerg, 55
Bougainville, 133, 134
Boundary Bay, 18
Bourg d'Ault, 76
Bournemouth, 165, 176
Bouvard Reef, 136
Bouvet I., 154
Brady Glacier, 15, 17
Brahmani R., 111
Brazos delta, 162
Brazos R., 33
Brazos Santiago Pass, 33
Bridgwater Bay, 74
Bridlington, 9, 60
Brigantine Inlet, 38
Brighton, 165

Bristol Channel, 4
Britannia Pt., 105, 166
British Honduras Reefs, 32
Brittany, 76, 77
Brookings, 20
Broome, 135
Broulee I., 167
Bruce Pt., 43
Brusand, 46
Bubiyan I., 109
Buchanan, 101
Budaki, 89, 167
Buenos Aires, 29
Buise, 57
Büjük Menderes R., 95
Bulacan R., 125
Bulbjerg, 55
Bulgaria, 87
Buna R., 84
Bunbury, 136
Burdekin delta, 146
Burlington Beach, 42
Burullos, L., 98
Bushire, 109
Butuan Bay, 126
Byobugaura, 120, 121, 165

Caba Roja, 32
Cabo Blanco, 27, 29
Cabo Frio, 30
Caboroan Spit, 125
Cabo San Juan, 103
Cabu I., 126
Cadiz, Gulf of, 78
Caguray R. delta, 126
Calais, 76
Calcasieu, L., 35
Calcasieu, R., 33
California, 20
California, Gulf of, 78
Callao, 27
Cambay, Gulf of, 110, 174
Cameroons, 103
Camiguin I., 126
Caminawit Spit, 126
Canala R., 151
Cananéia Outlet, 30
Canary Is., 156
Canche Estuary, 76
Canon beach, 20
Canterbury Bight, 150, 167
Caorle, 80
Cap d'Ailly, 4, 76
Cape Barrow, 14

Cape Bathurst, 41
Cape Billings, 124
Cape Blanc, 101
Cape Blanco, 21
Cape Bojador, 101
Cape Bon peninsula, 100
Cape Cantin, 101
Cape Coast, 105, 166
Cape Cod, 167
Cape Corrientes, 25
Cape Durrësit, 84
Cape Espenberg, 14
Cape Fear, 37
Cape Flattery, 19
Cape Glossa, 84
Cape Hangklip, 105
Cape Hatteras, 37, 38
Cape Henlopen, 38, 39, 167
Cape Intem, 126
Cape Kennedy, 37
Cape Kodori, 91
Cape Kolkas, 49
Cape Krusenstern, 14
Cape Kumuhaki, 153
Cape Leveque, 135
Capelhino, 154
Cape Lisburne, 13
Cape Lookout, 38
Cape Lopez, 103
Cape Mendocini, 20
Cape Mount, 101
Cape of Good Hope, 105
Cape Palliser, 149
Cape Pitsunda, 91, 167
Cape Prince of Wales, 14
Cape Recife, 105, 166
Cape Romain, 37
Cape Rozewie, 50
Cape St. Francis, 105
Cape San Blas, 35
Cape Sim, 101
Cape Shoalwater, 19
Cape Sukhumi, 91
Cape Tarafit, 101
Cape Thompson, 14
Cape Tryon, 40
Cape Turnagain, 149
Cape Valsch, 132
Cape Verde, 101
Cape Verde I., 156
Cape Vizcaino, 20
Cape Wom, 132, 166
Cape Woolamai, 142, 166
Cape Ferret Spit, 78

Captiva I., 36
Caraguatuba, 10, 30, 166
Caribbean Is., 36
Carmarthen Bay, 68
Carmel Bay, 21
Carmel, Mt., 97
Carolina, 38
Carpentaria, Gulf of, 134
Carson Inlet, 38
Carthage, 100
Casamance R., 101
Caspian Sea, 90–95, 160, 168, 171, 176
Castillos, 29, 165
Casuarina Pt., 145, 146
Cataisan Pt., 126
Catania, Gulf of, 81
Cathedral Rock, 24
Catumbela R., 103
Caucasus Ranges, 91
Cavite Spit, 125
Cayagan R., 125
Cayucos, 21
Ceyhan R., 162
Ceyhan delta, 96, 97
Chagos Bank, 156
Champon Bay, 116
Chañaral, 28, 163
Changi, 114
Changjiang estuary, 118
Channel Is. Habour, 23
Charlottetown, 41
Chatham, 40
Cheduba I., 113
Cheduba Strait, 113
Cheleken Peninsula, 92
Chenière au Tigre, 35
Cherbourg Peninsula, 76
Cervantes, 135
Chesapeake Bay, 38, 163
Chesil Beach, 71, 73
Chico R., 29
Chiloé, 28
Chisimaio, 108
Chittagong, 112
Chorokny R., 91
Christmas I., 156
Chukchi Sea, 14, 124
Cidurian delta, 128, 162
Cimanuk delta, 129
Cimanuk R., 130, 162
Cipunegara delta, 129
Citanduy, 131, 162
Citarum delta, 129
Clarence delta, 149

Clatsop spit, 19, 167
Cleeland Bight, 140, 142, 166
Cockburn Sound, 136
Cocos Is., 156
Coffins Bay, 137
Colombo, 111, 112
Colonia R., 30
Colonia Pt., 29
Colorado delta, 33
Colorado R., 25
Columbia R., 19, 162
Colville delta, 14
Comal delta, 130
Comoros archipelago, 156
Concepción, 28, 160
Conception Bay, 103
Conchagua Volcano, 26
Congo, 103
Constanza, 88
Constitución Cove, 28
Contarina, 80
Contas R., 30
Controller Bay, 15
Cook Inlet, 15, 17
Copper R. delta, 17, 160
Coquille R., 20
Corner Inlet, 140
Coromandel, 111
Coronado Strand, 25
Coronel, 28
Corsica, 80
Cosmoledo I., 156
Costa Rica, 26, 32
Cottesloe, 136
Cowes, 140
Cox's Bazaar, 112
Crete, 87
Crimea, 89
Cristobal Colon, 31
Cromer, 63
Cuba, 36
Cumberland, 74
Cypress Pt., 113
Cyprus, 97

Dahomey, 102
Damietta, 98
Danube delta, 88, 89
Danzig, Gulf of, 50
Dar-es-Salaam, 107
Darss foreland, 50, 167
Dasht R., 110
Davis Strait, 44, 160
Dawlish Warren, 176

Daytona Beach, 37
Dee estuary, 74
Deep Bay, 118
De Grey delta, 135
Den Helder, 57
Denison, 136
Deseado R., 29
Devils Slide, 21
Devon I., 41, 160
Devonport, 143
Disappointment, Cape, 19
Disko, 44
Djerba I., 100
Djibouti, 108
Djursland Peninsula, 52
Dnieper, 89, 163
Dniester, 89, 163
Dodanduwa, 112
Dog, I., 35
Don R., 89
Don delta, 91
Dondra Head, 111
Dorset, 64
Dothio R., 151
Drakes Bay, 21
Drin R., 84
Dubai, 109
Dutton Way, 171
Dumbéa delta, 151
Dume Pt., 166
Dun Aengus, 74
Dungeness, 68
Dungeness Spit, USA, 18
Dunwich, 64
Durban, 105, 166
Durban Bay, 106
Duren I., 108
Duxbury Pt., 21
Dymchurch, 175

East Anglia, 70, 71, 72
Easter I., 154
East Hampton, 38
East London, 105
Ebro delta, 79
Ediz Hook, 18, 19, 165, 175
Egypt, 108
Eighty Mile Beach, 135
Elburz Mountains, 92
Elis, 84
Elkhorn Slough, 21
Ellenbogen, 57
Ellesmere, L., 150, 167
Ellesmere, I., 42, 159

El Rompido, 78
El Salvador, 26, 28
Elson Lagoon, 13
Elwha R., 18, 19
Empire to Gulf of Mexico waterway, 35
Empress Augusta Bay, 133, 134
Encounter Bay, 137, 140
Endau R., 115
Ephesus, 95
Erdemli-Mersin, 97
Eregli, 95
Erie, L., 42
Erzen, R., 84
Esperance Bay, 137
Etang de Biguglia, 80
Ethiopia, 108
Evinas R., 84
Eyre Peninsula, 137
Eysk Peninsula, 89

Faeroes, 155
Fairy Dell, 64
Falcon Province, 31
Fal estuary, 74, 75
Falklands, 154
False Bay, 105
False Cape, 38, 166
Falstad, 46
Falsterbo, 46
Famagusta, 97
Fanø, 51, 52
Faraman, 79
Farewell Spit, 149
Far Pt., 68, 70
Fehmarn I., 50
Fernando de Noronha, 154
Fernando Po, 156
Figueira de Foz, 78
Filey Bay, 62
Filyos R., 95
Finland, Gulf of, 48, 49
Fire Island Inlet, 40
Flakket, 51, 167
Flanders coast, 76
Flaxman I., 13
Flinders I., 143
Fluvia R., 79
Flying Fish Pt., 145
Fonseca, Gulf of, 26
Formby, 68
Fort Bragg, 20
Fort Mendocino, 20
Foum el Oued Spit, 100
Foxe Basin, 42

Fraser delta, 18, 162
Fraser I., 145
Freetown, 101
Frisian Is., 57
Fuerte R., 25
Fuji R., 121
Funafati Atoll, 154, 155
Fundy, Bay of, 4, 40

Gaba estuary, 101
Gabes, Gulf of, 100
Gabo I., 140, 167
Galana R., 107
Galapagos, 154
Gallegos, R., 29
Galveston Pass, 33
Gambia R., 101
Gamtoos, 105
Ganges-Brahmaputra delta, 112
Ganges R., 111
Gascoyne R., 135
Gaza, 98
Georgia, 37
Georgia, Strait of, 18, 163
Geraldton, 135
Ghana, 101, 102
Ghar el Melh, 100
Gironde estuary, 78
Gisborne, 149
Glacier Bar, 16
Glenelg, 140
Godavari delta, 111
Godavari Pt., 167
Godavari R., 111
Goksu delta, 97
Golden Cap, 65, 66
Gold Coast, 145, 176
Goleta, 22
Gorgan Bay, 92
Gotland I, 46
Gough I., 154
Graham I., 18
Grand Bassam, 101
Grand Isle, 35
Grand Rhône R., 79
Grand Pacific delta, 16
Great Lakes, 42, 161, 169
Great Pearl Bank, 109
Great Yarmouth, 74
Greenland, 10
Greenville, 101
Grenen, 51
Grève de Goulven, 77
Greymouth, 149

Grijalva R., 32
Guadilquivir R., 78
Guajira Peninsula, 31
Guatemala, 25
Guayaquil, 27
Guayaquil, Gulf of, 27
Guilianova, 81, 167
Guinea, 101
Guinea Bissau, 101
Guinea, Gulf of, 101
Gunwalloe, 74
Guyot Glacier, 15, 16

Hab R., 110
Haifa, 97
Haifa Bay, 98
Hailing Is., 117
Hainan, 117
Half Moon Bay, 21, 171
Hallsands, 68
Hammamet, Gulf of, 100
Hangzhou Bay, 118
Hanjiang delta, 118
Hanstholm, 55
Hapuku R., 150
Hapuku delta, 150
Hatteras Inlet, 38
Havana, 36
Hawaii, 159, 169
Heard I., 151, 159
Heiligenhafen, 50
Helgenaes, 52, 161
Heligoland, 57, 159
Hel Spit, 50, 167
Heracleia, 75
Hereford Inlet, 38
Hermanus, 105
Hibok Volcano, 126
Hikada Range, 120
Hikkaduwa, 112
Hikman Peninsula, 108
Hilleh delta, 109
Hindenberg Dam, 57
Hindmarsh R., 140
Hirtshals, 55
Hispaniola Coast, 36
Hoed, 52, 53, 165
Hog I., 40
Hokkaido, 120, 123
Holderness, 60, 61, 62, 64
Holloways Beach, 145
Holy I., 72
Homer Spit, 14, 15
Honduras, 32

Hong Kong 117, 118, 166, 176
Honshu, 120
Hood Bay, 134
Hookena, 153
Hope R., 36
Hormuz, Strait of, 110
Hörnum Odde, 57
Horrocks Beach, 135
Huahine, 154
Huanghe R., 118, 162
Huarmey, 27
Hubbard Glacier, 15
Hudson Bay, 41
Humber estuary, 9
Humbug, Mt., 20
Hunts Bay, 36
Huon Peninsula, 132
Huon Gulf, 132
Huntingdon, 24
Huron, L., 42
Hurst Castle Spit, 167
Hutt delta, 147
Hvide Sande, 55, 56
Hyongsan Gang R., 119

Iberian Peninsula, 78
Iceland, 159
Icy Bay, 15, 16, 159
Idenao, 103
Ilha Grande, 30
Iloilo delta, 126
Indalsälven delta, 47, 171
Indramayu, 129
Indus delta, 110, 160
Inghuri R., 91
Ingolfshöfdi, 44
Inhambane, 106
Inman R., 140
Iohea, 134
Iquique, 28
Iraklion, 87
Irian Jaya, 132
Irish Sea, 60
Ir-Ramla, 83
Irrawaddy delta, 130, 162
Ishikari R., 120
Iskenderun, Gulf of, 97
Isla del Carmen 32
Isla Maydelena 25
Isla San Benedicto 25, 159
Isla San Lorenzo 27
Isle of Pines, 152
Israel, 108
Itanhém R., 30

Ivory Coast, 101
Ivo-Urika R., 134
Iwaki, 120

Jaba, 163
Jaba R., 133
Jaeren, 46
Jakarta, 130
Jakarta Bay, 129, 162
Jalauod delta, 126
Jamaica, 36
James Bay, 41
Japara, 130
Jask, 110
Jasons Bay, 115
Jatiluhur Dam, 129
Japan, Sea of, 122
Java, 127
Jequitinhonha, R., 30
Jiangsu, 118
Jogjakarta, 130
John Hopkins Glacier, 17
Johore, Straits of, 114
Jordan, 108
Joseph Bonaparte Gulf, 135
Juba R., 108
Jucururu R., 30
Juneau, 17, 159
Jurien, 135
Jutland, 51, 52, 159, 161, 167, 168

Kabylia, 100
Kachemak Bay, 14
Kagoshima Bay, 123
Kahoku-gata Lagoon, 122
Kahutara R., 150
Kaikoura Peninsula, 150
Kaipara Harbour, 147, 149
Kaitohe Beach, 141
Kakinada Spit, 111
Kalajoki, 48, 160
Kalimantan, 132
Kambara I., 152
Kamchatka, 123
Kanazawa, 122
Kara-Bogaz-Gol, 92, 93, 176
Karachi, 110
Kariba Dam, 107
Karmoy I., 46
Karoipa R., 151
Karpas Peninsula, 97
Kashima, 121
Katama Bay, 40
Kauai, 153

Kawakawa Bay, 149
Kawerong R., 134
Kedah R., 113
Kekaha, 153
Kelantan R. delta, 115
Kemp Welch R., 134
Kenai Peninsula, 14
Kenya, 107–108
Kerala, 111
Kerema Spit, 134
Khawr am Umayrah, 108
Khor Shori, 108
Kihola Bay, 153
Kilauea, 153
Kilia Subdelta, 88
Kilnsea, 62
Kimberley coast, 135
King I., 143
King Sound, 143
King William I., 41
Kiparissia, Bay of, 84
Kipini, 108
Kizilirmak R., 95
Klaeng, 116
Klaipeda, 50
Klaningrad, 50
Klim, 52
Klong Kas Po R., 119
Kodori R., 91
Køge Bay, 51
Kola Peninsula, 124
Kolobrzeg, 50, 167
Kolokhya I., 108
Kornat, 83
Kota Baru, 115
Kouaoua R., 151
Kouchibouguac Bay, 41
Koumac R., 151
Kowhai R., 150
Kowloon, 118
Kotzbue Sound, 14
Krakatau, 87, 127, 128, 159
Kranero, 87
Krasnovodsk Bay, 92
Kribi, 103
Krishna R. delta, 111
Kuala Dungari, 115
Kubai R. delta, 91
Kuban, 89
Kücük Menderes R., 95
Kucutari R., 115
Kudremalai, 111
Kuébéni R., 151
Kuiseb R., 103

Kujukurihama, 121, 167
Kukhtuy R., 124
Kunduchi, 107
Kura R., 92
Kura spit, 92
Kuria Muria Is., 108
Kurzeme Peninsula, 49
Kuskokwin R., 14
Kus Sla R., 117
Kutajara, 126
Kutch, Gulf of, 111
Kuwait, 109
Kwaipomata Pt., 134
Kwangtung, 118
Kyholm I., 55, 166
Kyushu, 122

Laccadive Is., 156
La Chaussée spit, 16
Lacosta I., 35, 166
La Coulée R., 151
Laem Pho spit, 116
Laem Sul, 116
Laem Talumphuk, 116
Laem Yai, 116
Laesø, 52, 54
Lagoa dos Patos, 30
Lagos, 102, 166
Laguna de Terminos, 32
Laguna Madre, 32
Lagune de San Rafael, 29, 159
La Jolla, 24
Lake County, 42
Lakemba I., 152
Lakes Entrance, 140, 143, 144, 167
Lake Worth Inlet, 39
Lakonikos Gulf, 87
La Ligua, Bay of, 28
La Masma, Gulf of, 78
Lambasa-Nggawa, 152
Lamington, Mount, 134, 161
Lamu, 108
Landes, 78
Langhkwai I., 113
Langlade I., 41
Langue de Barbarie, 101, 166
Lantau I., 118
Laptevykh, 124
La Punta spit, 27
Larap Bay, 125
Largs, 140
Laruma R., 134
Latium, 81
Latvia, 49

Legaspi, 123
La Havre, 76
Lema R. delta, 124
Lesha, 84
Lesser Antilles, 36
Levrier Bay, 101
Leyte, 126
Liberia, 101
Libreville, 103
Liguria, 80
Lighthouse Beach, 102
Lima, 27
Limantour spit, 21
Lim Chi Kong, 114
Limfjorden, 52, 55
Limpopo R., 106
Lincoln Beach, 35
Lingayan Gulf, 125
Lipari Is., 81, 101
Lithuania, 50
Lituya Bay, 15, 16
Lizard Peninsula, 74
Llantwit Major, 68
Llobregat R. delta, 79
Lobito, 103
Löderup, 46
Long Beach, 24
Long I., 38
Long Point, 42
Lonsdale, Point, 140
Lorne, 140
Los Angeles, 22
Los Angeles R., 24
Louisiana, 35
Lower Andalusia, 78
Loyalty Is., 152
Luan R., 118
Luanda spit, 103
Lüderitz, 165
Luleå, 46
Lummi R., 18
Luna, 125
Luria R., 107
Luzon, 125, 161
Lyme Bay, 65
Lyme Regis, 66, 68
Lyn R., 71
Lyngdal Fiord, 46
Lyngdalselva R., 46

Macajala Bay, 126
Machans Beach, 146
Mackenzie R., 42
Macquarie Harbour, 143

Madagascar, 107
Madang, 132
Madeira, 156
Madras, 111
Mae Nam Ta Pi, 116
Magdalen Is., 41
Maggona, 112
Maghna R., 112
Magilligan Foreland, 68, 167
Mahakam delta, 132
Mahanadi delta, 111
Mahanadi R., 111
Mahin, 102
Maine, 40
Makran, 160
Makran coast, 110
Makuluva cay, 152
Malago R., 126
Malaspina glacier, 115
Maldives, 156
Maliakos Gulf, 84, 86
Malindi, 107, 176
Malpleque, 41
Malta, 83
Mamberamo delta, 132
Mand delta, 109
Mandal, 46
Mandø I., 51
Mandurah, 136
Manganui Mt., 147
Mangatawhiri Spit, 147
Mangoky R., 107
Manila, 125
Manila Bay, 125
Manta, 27
Manukau Harbour, 147
Maputo, Bight of, 106
Mar Chiquita, 29
Mar del Plata, 29
Maracaibo, L., 31
Marambola R., 107
Mariager Fiord, 52
Marion Bay, 143
Marseilles, 79
Marthas Vineyard I., 40
Masirah I., 108
Mat R., 84
Matagorda I., 33
Matagorda Peninsula, 33
Matalscañas, 78, 167
Matrah, 109
Matsushima, 120
Maupiti, 154
Mauritania, 101

Mauritius, 156, 166
Mawizi I., 107
Mayo R., 25
Mayon Volcano, 125
Mayumba, 103
Mazagon 78, 167
Mazatlán, 25
Mba R., 152
Medanos Isthmus, 31
Medjerda, 162
Medjerda delta, 100
Medjerda R., 100
Mehrar delta, 109
Mekong R., 117
Melbourne, 163
Melen R., 95
Mellieha Bay, 83
Mellum, 57
Melville Bay, 43
Mendocino Fort, 20
Meob Bay, 103
Merang, 115
Merapi Volcano, 130
Mergui Archipelago, 113
Messenia, Gulf of, 84
Messier Fiord, 29
Metaponto coastal plain, 81
Mhlatuzi Lagoon, 106
Miami, 37, 166
Miani Lagoon, 110
Michigan, L., 42, 43
Middleton, 140
Miletus, Gulf of, 95
Miliani R., 100
Mindanao, 126, 161
Mindoro, 126
Miquelon, 41
Mira delta, 26
Miramar, 22
Miramar beach, 35
Miramichi Bay, 41
Mississippi Delta, 33, 35, 162
Mississippi subdeltas, 34
Miyazaki, 122
Moando, 103
Mobile Bay, 36
Mobile R., 36
Moçambique Channel, 156
Moila Pt., 134
Momi, 152
Mongpong delta, 126
Monomoy, 40
Monrovia, 101
Montague I., 14, 160

Monterey Bay, 21
Montevideo, 29
Mooréa, 154
Moreton I., 145
Morfa Dyffryn, 68
Morfa Harlech, 68
Morondava, 107
Morondava R., 107
Moroshechnaya R., 124
Morro Bay, 21, 167
Morro Rock, 22
Moruya, 143
Morzhovyets, I., 124
Mtunzini, 106
Mucuri R., 30
Muda R., 113
Mugu Lagoon, 23
Mui Bai Bung Peninsula, 117
Mullins Harbour, 134
Murray R., 140
Musandam Peninsula, 109
Mussulo, 103

Nador, 100
Nairn R., 71
Naka, 122
Nakéty R., 151
Namibia, 103
Nandi Bay, 152
Nandi delta, 152
Napier, 160
Narmada R., 111
Narva, 49
Nasser, L., 98
Nata, 31
Natori R., 120
Nauset Inlet, 40
Navarino I., 28
Navua delta, 152
Nayarit, 25
Ndakuinuku delta, 152
Ndawasamu delta, 152
Ndreketi delta, 152
Néaoua, 151
Negros I., 126
Népoui, 151, 163
Népoui delta, 151
Nestucca Spit, 19
Netanya, 98
Netherlands, 10, 58
Nethoni, 87
New Brunswick, 41
New Jersey, 38
New South Wales, 134–135

New Takoradi, 102
Newhaven, 142, 171
Newhaven Harbour, 74
Newlyn, 74
Newport, 19, 20, 167
Newport beach, 24
Nicaragua, 26, 32
Nicoya, Gulf of, 26
Niger delta, 102, 167
Niigata, 122
Nile delta, 98, 99, 163
Nile R., 97
Ninety Mile Beach, 140, 144, 169
Nishino-Shima Shinto, 123
Nissum Fiord, 55
Nooksack R., 18
Nordeney, 57
Norfolk, North, 63, 64, 72
Norris Glacier, 16
Northam, 68
Novaya Zemlya, 124
Nullabor Plain, 137
Nyali, 107
Nyiasia, 102
Nykarleby, 49
Nyong R., 103

Observation Pt., 140
Ocean Beach, 143
Ochamchire, 91
Ocracoke Inlet, 38
Odessa, 89, 167, 176
Odiel R., 78
Ofanto R., 87
Ogurchinsky, 92
Oi R., 121
Okains Bay, 150, 151
Okhota R., 124
Okhotsk, Sea of, 124
Olifants, 105
Oluanpi, 119
Oman, 108
Ombrone R., 81
Ono I., 35
Onslow, 135
Ontario, L., 42
Öraefajökull, 44
Orange R., 103, 105
Ord R., 135
Orford Ness, 68, 71, 166
Orinoco delta, 31
Orlowo Cliff, 50
Orantes delta, 97, 150
Oslo, 46

Oslo fiord, 46
Ostia, 81
Otway Sound, 28
Ouaco R., 151
Ouango R., 151
Oued Isser, 100
Ouha, 151
Ouinné R., 151
Oulu, 48
Outer Hebrides, 71

Pacific Pallisades, 23, 161
Padang, 126
Padre I., 33
Pahang delta, 115
Pajaro R., 21
Palawan, 126
Palembang, 126
Palisadoes, 36
Palmeirinhas Spit, 103
Palmyras Pt., 111
Palomino, 31
Palos Verdes Peninsula, 24
Pamisos R., delta, 84
Pampanga R., 125
Panama, 32
Panama, Gulf of, 26, 134
Panay, 126
Pancer Payang, 129
Panepane Pt., 147–148
Panguna, 134
Pantai Cinta Berahi, 115
Pantai Laut, 115
Papohaku Beach, 153
Par, 72, 163
Para R., 31
Paraguana Peninsula, 31
Paraíba do Sul Doce, 30
Parangtritis, 130
Paraná R., delta, 29
Pardo R., 30
Paria, Gulf of, 31
Patia R., delta, 26
Pasag R., delta, 125
Pasni, 110
Pattani, 116
Pattani R., delta, 116
Pattaya, 116
Peacock Spit, 19
Pearl R., 118
Pedro Pt., 112
Peleuz Spit, 77
Peloponnese, 84, 87
Pelican Pt., 103

Pella, 84, 85
Pelzerhaken, 50
Pelami R., delta, 130
Pembrey, 72
Penang I., 113
Pendine, 72
Peneus R., 84
Peniche, 78
Peñíscola, 79
Penn an C'hleuz, 77
Pensacola Bay, 35
Pensacola Beach, 35
Pensacola Inlet, 35
Pentewan, 72, 163
Peoples Republic of Yemen, 108
Perak R., 113
Perdido Bay, 35
Perpignan, 79
Perth, 136
Petit Rhône, 79
Peusangan R., 126
Phangna Bay, 112
Phillip I., 140, 141
Phillips Inlet, 35
Phuket, 113
Phu Quoc, 117
Piave R., 80
Picardy, 76
Pillar Pt., 21
Pinaroo Pt., 135
Piphot R., 117
Pirgos, 84
Pisa, 81
Pisco R., 27
Pismo Beach, 22
Plage d'Oué, 152
Planiski Channel, 83
Playa Caimanero, 25
Plenty, Bay of, 148, 161
Plettenburg Bay, 105
Po delta, 80
Point Barrow, 13, 14
Point Conception, 22
Point Delgado, 20
Point Dume, 23
Pointe aux Oiseaux, 101
Pointe aux Pins, 42
Pointe Banda, 103
Pointe d'Arcay, 77
Pointe de Beauduc, 78
Pointe de Grave, 78
Pointe de la Coubre, 78, 167
Pointe de la Gracieuse, 79
Pointe de l'Espiguette, 79

Pointe de l'Est, 41
Pointe de Pouthiauville, 76
Pointe de Tafe, 103
Pointe du Hourdel, 76
Pointe du Touquet, 76
Pointe Indienne, 103
Pointe Noire, 103
Point Franklin, 14
Point Hope, 14, 167
Point Lobos, 21
Point Loma, 24
Point Montara, 21
Point Mugu, 23
Point Negrais, 112
Point Pelée, 42
Point Reyes, 21
Point Riou, 15, 16
Point Sur, 21
Polihale Beach, 153
Pomerania, Gulf of, 50
Pomorie, 87
Ponce, 36
Pontchartrain, L., 35
Poole Harbour, 74
Poro R., 151
Portugese Bend, 24
Port Alfred, 105
Port Angeles, 19
Porta Pim, 134
Port Ballintrae, 76
Port Darwin, 134
Port Elizabeth, 105, 106, 166, 172
Port Fairy, 140
Porthallow, 74
Port Hedland, 135
Porthoustock, 74, 165
Port Heuneme, 22, 23
Portland, 140
Porto Alexandre, 103
Porto d'Ascodi, 81
Porto de Marco, 103
Porto di Chioggia, 80
Porto di Lido, 80
Porto di Malamocco, 80
Porto di Piave Vecchia, 80
Porto Vecchio, 80
Port Phillip Bay, 140
Portrush, 76
Port Stewart, 76
Potrerillos, 28
Poya R., 151
Praia de Fora, 30
Presque Isle Peninsula, 42
Prince Edward I., 40, 41

Prince Karl I., 46
Prince Rupert, 18
Princetown, 40
Prince William Sound, 17
Principe, 156
Provence, 166
Puerto Rico, 36
Puget Sound, 18, 163
Pulau Ular, 114
Punggol, 114
Punta Arena del Sur, 25
Punta Arenas, 26
Punta Chame, 26
Punta del Este, 29
Punta del Gato, 78
Punta Gorda, 20
Purari R., 134
Purbeck Peninsula, 68
Puttalam, 112

Qatar, 166
Qatar peninsula, 109
Qeshm I., 109
Queenscliff, 140
Queensland, 145, 146
Quintero, 167

Rabaul, 132
Ragged Pt., 21
Raiatéa, 154
Ramu R. delta, 132
Rangoon, 113
Rapid Bay, 140, 165
Ras al Hadd, 108
Ras al Mukalla, 108
Ras el Bar, 98
Ras el Bassit, 97
Ras Fortak, 108
Ras Hafun, 108
Rashid, 98
Ras Hilal, 99
Ras Kiromani, 107
Ras Madhroka, 108
Ras Ngomeni, 107
Ras Ormara, 110
Ras Rmel Spit, 100
Ras Tanurah, 109
Ras Umm Sa, 109
Rattray Head, 72
Ravda, 87
Ravenna, 80
Rayong, 116
Redondo Beach, 24
Refugio State Park, 22

Restinga de Marambaia, 30
Réunion, 156
Rewa R., 152
Rharb Plain, 101
Rheban Spit, 143
Rhine R., delta, 59
Rhode I., 40, 174
Rhône R., delta, 79, 163
Ria del Eo, 78
Richards Bay, 106
Riga, Gulf of, 49
Rimini, 83
Ringköbing fiord, 55, 56
Rio Camana delta, 27
Rio Colorado delta, 29
Rio de Janeiro, 30
Rio de la Plata, 29, 168
Rio del Oro, 101
Rio Goascoran, 26
Rio Grande, 33
Rio Grande de Santiago, 25, 162
Rio Grande de San Miguel, 26
Rio Grande do Sul, 30
Rio Jiboa, 26
Rio Ligua, 28
Rio Maipo, 28
Rio Mitare delta, 31
Rio Negro delta, 39
Rio Petorea, 28
Rio Rioni, 91
Rio Rimac, 27
Rio Samala, 26, 161
Rio Sinu, 32, 162
Rio Tambo delta, 27
Rio Tumbes, 27
Rio Uruguay, 29
Roanoke I., 4
Rodriguez I., 156
Rockall I., 156
Rogue R., 20, 167
Røjle Klint, 52
Rømø I., 52
Rosas, Gulf of, 79
Rose Pt., 18
Rosetta estuary, 98
Rosnaes, 52, 161
Rosslare Harbour, 76
Rotes Kliff, 51
Rovuma R., 107
Ruan Lanihorne, 74
Rubjerg, 55
Rügen, 50
Rutherfords Beach, 35

Sabine Pass, 33, 35

Sable Cape, 36
Sable I., 155
Sacramento R., 21
San Francisco Bay, 21, 163
Sagami Bay, 121
Saguarema, 30
Saharan Coast, 101
St. Georges I., 35, 36
St. Helena, 154
St. Helena Bay, 105
St. Ives Bay, 72
St. Josephs Point, 35
St. Kilda I., 156
St. Lawrence estuary, 41
St. Lawrence, Gulf of, 40, 41
St. Lucia lagoon, 106
St. Malo, Bay of, 77
St. Pauls Bay, 83
St. Pierre I., 41
St. Vincent Gulf, 137
St. Vincent I., 35, 36
Sajid Rud R., 92
Sakarya R., 95
Sakhalin, 123, 124
Salado R., 28
Saldanha, 105
Salina Cruz, 25
Salinas, 27
Salinas R., 21
Salonika, 84
Sal Pt., 22
Salvador, 31
Samarai, 134
Sambian Peninsula, 42
Samoa, 154
Samur delta, 92
Sanaga R., 103
San Antonio, 28
San Augustin, Cape, 126
San Diego Bay, 24
Sandoway R., 113
Sand Pt., 20
Sands of Forvie, 72
Sandy Hook, 38, 175
Sandyland, 22
Sandy Pt., 3, 140
Sandwich Harbour, 103
Sandwip, 112
San Gabriel R., 24
Sanggarung delta, 130
Sanibel I., 36
San Joacquin R., 21
San Jorge, Gulf of, 29
San Juan delta, 26
San Luis Pt., 22

San Matias, Gulf of, 29
San Miguel I., 22, 165
San Nicolas I., 22
San Pedro Bay, 24
San Pedro Pt., 21
Santa Ana R., 24
Santa Barbara, 22
Santa Catalina I., 22
Santa Catarina State, 30
Santa Clara, 23, 24
Santa Cruz, 21
Santa Maria Cape, 106
Santa Maria R., 21
Santa Monica, 24
Santa Monica Mts., 23
Santa Monica Volcano, 25
Santander Bay, 78
Santa Rosa I., 35
Santa Ynez R., 21
Santoft Beach, 147
Santorini, 161
Santos, 30
São Francisco R., 30
São Mateus R., 30
São Tomé, 156
Sapporo, 120
Saquisi R., delta, 126
Saraforo, 87
Saraoui, 101
Sardinia, 80, 81
Sars Spit, 46
Sassandra, 101
Satun, 113
Saua R., 134
Saudi Arabia, 108
Savanna R., 37
Scarborough Beach, 136
Scarborough Bluffs, 43
Scheveningen, 59
Schlei R., 50
Scott Head I., 59
Seaside, 19, 176
Sebastian Vizcaino Bay, 25
Sechura Desert, 27
Sedili Kecil, 115
Seenigama, 112
Segara Anakan, 130, 131, 162
Selvagens Archipelago, 156
Seletar, 114
Sellar Pt., 143
Seman R., 84
Sembawang, 114
Semenovskii I., 124
Senegal R., 101
Sendai Bay, 120

Sengeyski Spit, 124
Sepik Delta, 160
Sepik R., 132
Serchio R., 81
Seven Mile Beach, NSW, 145, 167
Seven Mile Beach, Tasmania, 143
Seychelles, 156
Seyhan-Ceyhan, 97
Seyhan delta, 97
Shanghai, 118
Shannon Pt., 18
Shantung Peninsula, 118
Sheena R., 18
Shell Beach, 21
Shell I. Spit, 35
Sherbro I., 101
Sheringham, 63
Shinano R., 122, 162
Shiretoko Peninsula, 123
Shkumbin R., 84
Shoalhaven R., 145
Shortland I., 134
Siam, Gulf of, 116
Sibalom R., 126
Siberia, 159
Sicily, 80, 81
Sidi Fredj, 100
Sidi Youcef, 100
Sierra Leone, 101
Siletz R., 19
Sinaloa, 25
Singapore I., 113, 114
Singkel Plain, 126
Sirbangis, 126
Sirte, Gulf of, 99
Sissano Lagoon, 132, 161
Sjerø Bay, 51
Skåne, 46
Skull Cliff, 14
Skyrig Sound, 28
Sochi, 176
Socotra, 108
Sofala, Bight of, 106
Solent Region, 74
Solo delta, 130
Solomon Is., 154
Somerset I., 41, 160
Somme estuary, 76, 77
Song Hong R., 117
Sonmiani, 110
Sossus R., 103
Souelaba Pt. Spit, 103
Sous Plain, 101
South Africa, 105–6

South China Sea, 116
Southampton Water, 74
Southport, 68
Southwest Pass, Mississippi delta, 33
Southwold, 71
Spencer Gulf, 137
Spencer Pt., 14
Sperchiós R., 84
Spey R., 71
Spitsbergen, 46
Spoegrivier, 105
Split, 83, 84
Spurn Head, 9, 62, 63
Stanwell Park, 145, 173
Starigrad, 83
Stavanger, 46
Stilbaai, 105
Stradbroke I., 145
Streaky Bay, 136, 137
Stromboli I., 81
Stuart Pt., 134, 135
Sudan, 108
Suffolk, 9, 64
Suislaw R., 20
Sukhumi, 167
Sukhumi Bay, 91
Sulawesi, 132
Sulina Breakwater, 88
Sumatra, 126, 127
Sumba, 127
Sundays R., 105
Sungai Jemaluang, 115
Sungai Sedili Besar, 115
Surat Thani, 116
Surtsey, 44, 45, 123, 159, 161
Susitna, 17
Suwannee R., 36
Suva, 152
Svealand, 47
Swakop R., 103
Swan I., 140
Swina Inlet, 50
Szczecin, L., 50

Tabasco, 32
Tagliamento R. delta, 80
Tahiti, 154
Tahunanui spit, 150
Tainan, 119
Taipei, 119
Taiwan, 119
Takanabe, 132
Taketomy, 124
Talcahuano, 28, 160

Tallinn, 49
Tampa Bay, 35
Tana R., 107–108
Tangalla, 111
Tanzania, 107
Tapti R., 111
Taranaki, 149
Taranto, Gulf of, 81
Tarawera Volcano, 148
Tarokina R., 134
Tasmania, 140
Tauranga Habour, 148, 171
Taylor Bay, 15, 17
Taymyr Peninsula, 124
Tenasserim, 113, 163
Tendra Spit, 89
Tendrovskaya Kosa, 167
Tenryu R., 121
Tenshaw R., 36
Tentsmuir Pt., 72
Tepic, 25
Terek R. delta, 92
Terembu Retan Laut, 114
Terengganu R., 115
Tevere R., 92
Thames estuary, 72
Thera I., 87
Therasia I., 87
Thermaïkós Bay, 84
Thermopylae, Pass of, 84
Thio. R. delta, 151, 163
Tiber R. delta, 81
Tierra del Feugo, 29
Tigris–Euphrates–Karun delta, 109
Tillamook Bay, 20
Tillamook Head, 20, 161
Timaru, 150
Timmendorf, 50
Tinip R., 151
Tipaza, 100
Tobruk, 99
Togo, 102
Tokai, 121
Tokyo, 121, 160, 176
Tomakomai, 120
Tomales Bay, 20, 163
Tonkin, Gulf of, 117
Torre Astura, 81
Torres, 30
Torricelli Ranges, 132
Torsminde, 55
Tosa Bay, 122
Toyama Bay, 122
Tres Maria Is., 25

Trinchera-Quivologo, 28
Trincomalee, 111
Trindade, 154
Trinidad, 37
Tripoli, 99
Tristan da Cunha, 154
Tronto R., 81
Trous Sans Fond, 101
Truando R., delta, 31
Trujillo, 27
Tsondab R., 103
Tsiribihina R., 107
Tuapse R., 91
Tubingan Pt., 126
Tucacas, 31
Tuléar, 107
Tumaco, 26
Tumbes R., delta, 27
Tumpat Bay, 115
Tunis, 100
Turnagain Heights, 14, 161
Turner Glacier, 15
Turners Peninsula, 101
Tweed Heads, 143
Twilight Cove, 4, 137
Tyub-Karagan Peninsula, 92

Ulhas R., 111
Ulverstone, 143
Umeå, 46
Umm Said Sebkha, 109
Umuiden, 57, 59, 167
Unonio Bay, 107
Uraba, Gulf of, 31
Utique, 100

Vaasa, 47
Vadehavet, 52
Vaigatch I., 124
Vailala G., 134
Valdez Glacier, 17, 161
Valdez Peninsula, 29
Valparaíso, 28
Vancouver I., 18
Van Diemens Gulf, 134
Vanua Levu, 152
Varde, 52
Varna, Gulf of, 87
Varoi R., 134
Var R., 79
Vassilevskii, 124
Venezuela, Gulf of, 31
Venice, 80, 160, 175
Ventura R., 22

Veracruz, 23
Vermilion Bay, 33
Victor Harbour, 140
Victoria, Australia, 138–140
Victoria Beach, 102
Victoria I., 41
Victoria R., 135
Vigsø, 55
Vilamoura, 78
Vilanova i la Geltrú, 79
Visakhapatnam, 111
Vistula R., 50
Viti Levu, 152
Voise R., 84
Volga R. delta, 102, 163
Volturno R., 81

Wadden Sea, 57
Waikiki Beach, 153, 176
Wailevu R. delta, 152
Waimangaroa, 149
Waimbula R., delta, 152
Waimea Bay, 153
Waini R., 31
Wainkoro R. delta, 152
Wainwright, 14
Wairepo Lagoon, 150
Walvis Bay, 103
Wanganui Harbour, 147
Warrnambool, 140
Wasaga Beach, 42
Wash, The, 59
Wasque, 40
Wedge, I., 135
Wellington, 147, 148, 160
Wenduine, 59
West Bay, 74, 165
Westernport Bay, 140

Westham I., 18
Westland, 149
Wewak, 132
Wexford, 76, 166
Weymouth, Tasmania, 143
Wilhelmshaven, 57
Willapa Bay, 19
Wilsons Promontory, 140
Windar R., 110
Winterton Ness, 68, 167
Wladyslawowo, 50
Wonnerup, 136
Wonsan Bay, 119
Workington, 163
Wrangel I., 134
Wulun Canal, 130

Yachats, 20
Yakatut Bay, 15
Yanakie Isthmus, 140, 166
Yaquina Bay, 19
Yellow Sea, 118
Yemen, 108
Yeongil Man Bay, 119
Yesilirmak R., 95
Yilan Bay, 119
Yokohama, 121
Yorkshire, 9
Yoshino R., 122
Ystad, 46
Yucatan Peninsula, 32
Yukon R., 14
Yumiga-Hama, 122

Zambezi delta, 106, 163
Zeebrugge, 59
Zeila, 108
Zhujiang delta, 117
Zohreh delta, 109

Subject Index

Analysis of coastline changes, 3
Archaeological sites, 2, 74, 76, 79–80,
  84–87, 95, 99–100, 109, 136
Army Corps of Engineers (USA), 10, 15
Artificial beach nourishment, *see* Beach
  nourishment, artificial

Barrier (Barrier Island), 13, 14, 25, 30,
  32, 33, 35, 36, 37, 38, 40, 41, 42,
  49, 51, 52, 55, 57, 69, 73, 76, 81,
  84, 88, 89, 90, 92, 97, 101, 105,
  109, 111, 112, 119, 120, 123–124,
  135, 146–147, 160, 161, 168–69
Beach depletion, due to increased
  storminess, 172–173
  due to rising water-table, 173
  due to volume reduction, 172–173
Beach nourishment, artificial, 19, 25, 38,
  42, 59, 89, 91, 105, 114, 141, 145,
  153, 163–168, 176
  from coastal quarrying, 52–53, 74,
    113, 140
  from cliff erosion, 19, 21, 29, 40, 52,
    71, 84, 120
  from dune spilling, 22, 27, 29,
    101–103, 105, 109, 142, 165–166
  from rivers, 10, 19, 21, 24, 28, 30, 35,
    36, 38, 47, 52, 72, 74, 81, 87, 95,
    97, 98–99, 103, 105, 107, 112, 113,
    115, 122, 125, 127, 128–130, 131,
    132–134, 140, 145, 147, 151, 164,
    166, 168, 172
  from sea floor, 29, 31, 37, 38, 46, 48,
    52, 55, 68, 70, 72, 84, 88, 92, 100,
    103, 105, 108, 109–110, 112, 114,
    116, 118, 125, 136, 137, 144, 147,
    150, 151–152, 156, 160, 166, 168
Beach mining, 19, 36, 37, 68, 74, 76–77,
  91, 97, 98, 100, 103–105, 112–113,
  153, 165, 171
Beach lobe, 14

Beach progradation, by convergent
  drifting, 33, 46, 81, 121, 143
Beach Protection Authority
  (Queensland), 145
Beach ridges, 19, 25, 30–33, 37, 40, 43,
  48, 52–53, 84, 97, 101–102, 107,
  115–116, 123–126, 128, 133–134,
  147, 154–155
Beach rock, 27, 37, 97, 102, 113, 116,
  135, 156
Beagle, The, 28
Biogenic sediments, 32, 89, 92, 98, 109,
  152–153, 168
Blockhouses, undermined by erosion,
  28, 55, 78
Boulder armouring, 19, 38, 112, 175
Breakwater, 19, 20, 21, 22, 23, 24, 27,
  28, 32, 33, 35, 36, 37, 38, 40, 42,
  50, 52, 55, 56, 59, 74, 78, 79, 80,
  81, 88, 91, 98, 99, 100, 101, 102,
  105, 106, 109, 111, 114, 121, 135,
  140–141, 143, 144, 147, 149, 150,
  171, 172, 176
Bremen, Adam von, writing on
  Heligoland, 57
Bruun theory, 43, 169–170

Causeway, 29, 36, 57, 117
Cay, 32, 130, 146, 152, 154–155, 156
Cliff recession, 20, 22, 24, 25, 31, 40,
  43, 45, 50, 52, 55, 59–68, 76, 80,
  81, 83, 84, 87, 89–90, 97, 99–100,
  105, 107, 109, 111, 114, 117, 119,
  122, 124, 127, 134, 137, 140, 149,
  152, 154, 156–157, 158–159
Coast and Geodetic Survey (USA), 14
Coastal Erosion Information System
  (USA), 10
Coastal Studies Institute, Louisiana
  (USA), 12
Coastline, 4–5

Commission on the Coastal Environment (IGU), 1, 11, 167–168
Coode, Sir John, 1852 survey of Chesil Beach, 73
Crown-of-thorns seastar, 146
Cuspate foreland (Cuspate spit), 14, 18, 20, 23, 25, 35, 37, 44–45, 51, 53, 68, 83, 91, 103, 108, 118, 128, 133, 134, 136–137, 143, 147, 153
Cyclone, 111, 135, 152, 154, 156. See also Hurricane, Typhoon

Darwin, Charles, observed Talcahuano earthquake, 28
De La Favolière's Survey (France), 76
Delta, 14, 17, 18, 20, 25, 26, 28, 29, 30, 31, 32, 33, 35, 36, 41, 42, 44, 79, 80, 81, 84, 88–91, 92, 94, 95, 96–97, 102, 103, 106–107, 109–111, 112, 113, 115, 116, 117, 118, 120, 122, 124, 126, 128–132, 135, 146–147, 149, 150, 151, 152, 160, 161, 172
Deltaic fan, 20, 22, 71, 83, 87
Depositional capes, 13, 83
Des Barres' Chart (Massachusetts), 40
Dunes, cliffed, 32, 35, 52, 55, 76, 111, 135, 140, 144, 156
Driftwood, effects of, 20, 124
Drumlin, 40, 59

Earthquakes, 14–15, 17, 26, 28, 87, 97, 110, 112–113, 121, 125, 132, 140, 147–148, 152, 154, 160–161
Eduard Bohlen, shipwreck in Namibia, 103
Emerged (Emerging) coastlines, 14, 17, 28, 46–49, 87, 91–95, 97, 110, 113, 121, 132, 160–161

Fiord (Firth, Loch), 15–18, 28, 29, 42, 44, 46, 72, 156
Fishponds, 117, 125, 174–175
Fusilier, shipwreck in New Zealand, 147

Geodetic Institute, Copenhagen (Denmark), 51
Geographical Survey Institute (Japan), 119
Glacial activity, 15, 28, 43–44, 149, 157, 159–160, 161
Glacial outwash plain (Sandur), 17, 44, 149, 157, 168

Groyne, 9, 19, 38, 42, 57, 59, 63, 81, 112, 114, 120, 122, 140–141

Headlands, eroded to islands, 112
Historical photographs, 4, 7, 20, 22, 24–28, 36, 37, 40, 41, 77, 122, 137, 145
Historical records, 4, 84–87, 100
Holocene marine transgression, 3, 60–61, 63, 72, 89, 112, 116, 123, 173
Hurricane, 32, 33, 36, 37. See also Cyclone, Typhoon

Ice coast, 15–17, 28–29, 41–42, 43–44, 124, 154, 156–157, 159–160
Instituto Hidrografico de la Marina (Spain), 78
Inter-tidal zone, 5
Isostatic movements, 17, 42, 44–49, 52–53, 160
Istituto Geografico Militare (Italy), 80

Lagoon Entrance (Tidal Inlet), 33, 35, 38, 40, 57, 78, 80, 105, 106, 137, 140
Landsat, 7, 112
Landslide, 20, 21, 23, 43, 52, 63–67, 83, 87, 89, 91, 109, 126, 132, 161–162
Liddell's Survey (Jamaica), 36

Machair, 72
Mangrove, 26, 27, 31, 32, 33, 36, 101, 103, 106, 108, 109, 110, 112, 113, 114, 116, 117, 118, 125, 126, 130, 131, 132, 134, 135, 137, 140, 147, 152, 156, 160, 166–167, 174–175
Marco Polo, visit to Palembang, Sumatra, 126
Measurement of coastline change, 3, 5, 6, 9, 10–11
Mud volcano, 113

National Shoreline Study (USA), 10–11, 35, 42
Nearshore dredging, effects of, 171
Nero's Villa, Anzio, 81–82

Offshore breakwater, 23 83, 98
Ord, 62, 159

Periglacial activity, 13–14, 41, 64, 76, 124, 154, 157, 159–160
Pixel, 7

Posidonia, 79

Ra moraine, 40
Rates of coastline change, 3, 10–11, 158
Reclamation, 9, 21, 52, 57, 59, 78,
    113–114, 117–118, 119, 121, 136,
    160, 176
River outlet diverted, by canal, 33, 35,
    81, 91, 122, 128, 130, 162
  by flood, 22, 24, 32, 92, 96–97, 100,
    118, 126, 129–130, 162
  by log jam, 18
River sediment yield diminished, by
    catchment bedrock exposure, 87,
    97, 163
  by channel dredging, 81, 122, 127, 163
  by conservation works, 81, 95, 163
  by dam construction, 19, 24, 35, 47,
    77, 79, 81, 89, 91, 98–99, 106–107,
    122, 129, 163, 171
River sediment yield increased, by
    hinterland quarrying, 21, 28, 72,
    74, 103, 105, 113, 115, 125,
    132–134, 151, 163
  by soil erosion, 28, 36, 38, 81, 87, 95,
    112, 128–130, 131, 140, 147, 162
River sediment yield modified, by
    climatic variations, 10, 24, 30, 81,
    87, 107, 145
Royal Commission on Coast Erosion
    (Britain), 8–10, 59
Russian–American Company, 15

Salt marsh, 5, 18, 21, 29, 38, 41, 46, 48,
    52, 54, 57, 74–75, 77, 80, 110, 118,
    119, 138–139, 160, 174–175
Sandsteilkante (Beach scarp), 26
Sandur, *see* Glacial outwash plain
Satellite imagery, 7, 8, 112
Sea floor deepening, 46, 62, 122
Sea level changes, 161, 169–170
Seaside resorts, 19, 107, 112, 145, 154,
    176
Sea wall, 9, 19, 38, 42, 57, 59, 63, 81,
    112, 114, 120, 122, 140–141, 159,
    165, 171, 175–176
Sebkha, 99, 109
Sediment budget, 5, 68–71, 174
Sharm, 108
Shipwreck, 28, 91, 103, 137, 147, 167
Shoreline, 4–5
Silt jetties, 33
Spartina, 74, 118

Spit, 14–15, 18–20, 22, 23, 24, 26, 27,
    29, 31, 35, 36, 37, 38, 39, 40, 41,
    42, 46, 48, 50, 51, 57, 59, 62, 68–70,
    76–78, 79, 80, 81, 84, 88, 89, 92,
    98, 100, 101, 103, 107, 108, 109,
    111, 112, 113, 114, 115–116, 118,
    119, 120, 124, 125, 126, 133, 134,
    135, 136, 137, 140, 145, 146–147,
    149–150, 151, 154, 157, 160, 163,
    166, 167, 175–176
Stack, 20, 22, 55
Stone reef, 31
Storm surge, 4, 41, 42, 57, 68–69, 112,
    117, 118, 152, 166, 169
Striations, 45, 49
Submarine canyon, 21, 23, 101
Submarine slumping, 17, 46, 101, 161
Submerged (Submerging) coastlines, 14,
    15, 17, 26, 28, 110, 132, 160–161

Tarawere explosion, 161
Tebenkov Atlas, 15
Tectonic subsidence, 14–15, 17, 26, 28,
    109, 110, 132, 160–161
Tectonic uplift, 14, 17, 28, 46–49, 87,
    91–95, 97, 110, 113, 119, 121, 132,
    147, 160–161
Tetrapods, 91, 105–106, 109, 122–123,
    176
Tidal inlet, *see* Lagoon Entrance
Tide range, 4, 29
Tide records, 41, 72, 74, 160, 170–171
Tombolo, 22, 29, 41, 59, 79, 84, 87, 98,
    110, 119, 124, 140, 167
Towns stranded inland by deposition,
    84, 95, 110
Tsunami, 4, 15–16, 28, 87, 128, 153
Tundra bluff, 13, 14, 159
Typhoon, 117, 118 *See also* Cyclone,
    Hurricane

Uffico Idrafico della Marina (Italy), 80

Vancouver's chart, 16
Volcanic activity, 14, 25–26, 44–45, 81,
    87, 108, 122–123, 124–125,
    126–128, 130, 132–134, 147–148,
    152–154, 157, 159, 161

Working Group on Coastline Changes
    (IGU), 1, 11

Zeegat, 57
Zostera, 57